Antimicrobial Stewardship in Non-Traditional Settings

Shira Doron • Maureen Campion
Editors

Antimicrobial Stewardship in Non-Traditional Settings

A Practical Guide

Springer

Editors
Shira Doron
Division of Geographic Medicine
and Infectious Diseases
Tufts Medical Center
Boston, MA, USA

Maureen Campion
Department of Pharmacy
Tufts Medical Center
Boston, MA, USA

ISBN 978-3-031-21709-8 ISBN 978-3-031-21710-4 (eBook)
https://doi.org/10.1007/978-3-031-21710-4

This Springer imprint is published by the registered company Springer Nature Switzerland AG
The registered company address is: Gewerbestrasse 11, 6330 Cham, Switzerland

Introduction

Antimicrobials are critical life-saving tools of modern medicine, but they are also dangerous. Rapid drug development in the early days of antibiotics meant that even as resistance was developing, prescribers could almost always count on a new antimicrobial coming to market that would treat the increasingly resistant pathogens. But the drug development pipeline is no longer keeping up with the growing resistance trend. As a result, a "post-antibiotic era," where minor infections could once again kill and advanced therapies like cancer chemotherapy and organ transplantation could prove too dangerous, looms large.

Antimicrobial stewardship (AS) is the coordinated effort of measuring and improving the way antimicrobials are prescribed and used, with the goal of maximizing benefit and minimizing harms. Those harms include not only the promotion of antimicrobial resistance but also unnecessary adverse events, side effects, and costs. Antimicrobial stewardship is critical to preserving our ability to use antimicrobials in the future.

In 2014, CDC called on acute care hospitals to implement antimicrobial stewardship programs based on the *Core Elements of Hospital Antibiotic Stewardship Programs* [1]. In 2015, the *Core Elements of Antibiotic Stewardship for Nursing Homes* [2] were released. In 2016, the *Core Elements of Outpatient Antibiotic Stewardship* [3] were launched, with the intent to encompass clinics, emergency departments, dental offices, and urgent care centers. The goal is to continue to expand stewardship efforts to every corner of health care in which antimicrobials are in use. However, the implementation of antimicrobial stewardship requires resources—including time, clinical expertise, and compensation.

The 2007 Infectious Diseases Society of America consensus guidelines recommended that AS programs be established with the dedicated, compensated time of Infectious Disease specialty trained physicians and pharmacists [4]. However, many healthcare settings lack the resources necessary to adhere to this recommendation. Nevertheless, AS can be effectively adapted to environments with limited infectious diseases resources.

In this book, we provide in-depth reviews on approaches to antimicrobial stewardship, from the implementation of successful initiatives to the development of formal programs, in non-traditional, often resource-limited, settings. The authors, who are physicians, clinical pharmacists, and veterinarians, are experts who practice in those settings and have worked therein to improve the prudent use of antimicrobials. This information will hopefully help health care professionals to initiate or enhance their antimicrobial stewardship practices.

Acknowledgment: We thank Dr. Kirthana Beaulac for her work on the conception of this book and early editing.

References

1. CDC Core Elements of Hospital Antibiotic Stewardship Programs. https://www.cdc.gov/antibiotic-use/core-elements/hospital.html.
2. CDC Core Elements of Antibiotic Stewardship for Nursing Homes. https://www.cdc.gov/antibiotic-use/core-elements/nursing-homes.html.
3. CDC Core Elements of Outpatient Antibiotic Stewardship. https://www.cdc.gov/antibiotic-use/core-elements/outpatient.html.
4. Dellit TH, Owens RC, McGowan JE Jr, et al. Infectious Diseases Society of America and the Society for Healthcare Epidemiology of America guidelines for developing an institutional program to enhance antimicrobial stewardship. Clin Infect Dis. 2007;44(2):159–77.

Division of Geographic Medicine and Infectious Diseases — Shira Doron
Tufts Medical Center
Boston, MA, USA
e-mail: sdoron@tuftsmedicalcenter.org

Department of Pharmacy — Maureen Campion
Tufts Medical Center
Boston, MA, USA
e-mail: maureen.campion@tuftsmedicine.org

Contents

Antimicrobial Stewardship in Long-Term Care Facilities (Nursing Homes and Rehabilitation Centers)

1

Lucy S. Witt, Sheena Kandiah, and Sujit Suchindran

Introduction

Antimicrobial Stewardship Programs and Why Your Facility Should Have One

Antimicrobial overuse is a well-described problem in long-term care settings such as nursing homes and sub-acute rehabilitation centers [1–3] and is quickly becoming more recognized in acute rehabilitation centers [4, 5]. An antimicrobial stewardship program (ASP) is an institutional program aimed at optimizing antimicrobial usage with the goal of improving patients' health by reducing the number of healthcare associated infections such as *Clostridioides difficile* infections (CDI) and multi-drug resistant organisms (MDRO) along with medication-associated adverse events. In 2017 the Centers for Medicare and Medicaid Services (CMS), the federal agency that determines Medicare and Medicaid regulations, required that all participating long-term care facilities (LTCFs) and rehabilitation centers establish an ASP that must include "antibiotic use protocols and a system to monitor antibiotic use" [6]. A survey conducted in 2018 showed that a majority of US nursing homes had at least a "moderately comprehensive" ASP program, an increase from prior studies [7]. For the purposes of this chapter the term rehabilitation centers will be used to describe both acute and sub-acute rehabilitation facilities. The term long-term care facilities will be used to describe institutions where residents reside indefinitely.

Numerous studies have described antibiotic overuse and its relationship to increased *CDI,* MDROs, and adverse medication effects [2, 3, 8]. Antibiotic overuse and misuse lead to excess cost to a facility from the costs of the unnecessary

L. S. Witt · S. Kandiah · S. Suchindran (✉)
Division of Infectious Diseases, Emory University School of Medicine, Atlanta, GA, USA
e-mail: lucy.s.witt@emory.edu; sheetal.kandiah@emory.edu; sujit.suchindran@emory.edu

S. Doron, M. Campion (eds.), *Antimicrobial Stewardship in Non-Traditional Settings*, https://doi.org/10.1007/978-3-031-21710-4_1

antibiotics themselves, the costs of using expensive antibiotics when less expensive ones could be substituted, and from treatment of downstream MDROs or medication side effects as well as increased length of stay in acute care facilities. ASPs have been shown to alleviate some of this financial burden [9–11]. Furthermore, residents living in long-term care facilities with inappropriate antibiotic use are at higher risk for adverse outcomes such as CDI, MDROs, and medication side effects regardless of whether or not they receive an antibiotic prescription [3]. A majority of nursing home residents will receive an antimicrobial in the course of a year [1, 12] and 40–75% of these prescriptions may be inappropriate [2] (Fig. 1.1). Residents of long-term care facilities and sub-acute rehabilitation centers may be frail and prone to infection making it challenging for providers to balance between caring for these patients while avoiding antibiotic overuse. Similarly, infections are known to be a major cause of nursing home resident morbidity and mortality and are the most common reason for transfer from a nursing home to an acute care hospital [1]. There is significantly more variability in antimicrobial usage between LTCFs compared to acute care hospitals making these LTCFs an important target of antimicrobial stewardship interventions [3, 13]. In this chapter we aim to summarize and expand upon the importance of ASP, describe practical aspects of ASP implementation, and review existing literature supporting certain interventions. Throughout we will highlight specific areas of focus that may be applicable in LTCFs and rehabilitation centers.

Acute Rehabilitation Centers, Sub-Acute Rehabilitation, and Long-Term Care Facilities

We acknowledge that antimicrobial stewardship strategies may differ depending on the type of rehabilitation setting. Anticipating each institution's stewardship needs requires an understanding of the differences between the settings. Acute rehabilitation facilities care for patients who can be expected to make substantial improvements in functional status over a shorter period of time. They can be a free-standing

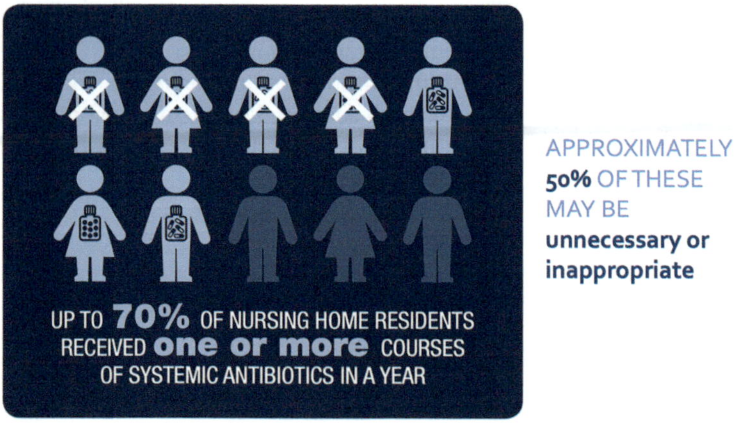

Fig. 1.1 Antibiotic use in long-term care facilities. Adapted from CDC Core Elements

hospital or housed within an acute care hospital. Often acute rehabilitation patients also require ongoing medical treatment in combination with physical, occupational, or speech therapy. At these facilities, clinical care and leadership of a multidisciplinary care team is provided by physicians trained in rehabilitation medicine who provide daily face-to-face patient assessments. A physician is available at all times for emergencies. These patients also receive 24-h nursing care and more than three hours of therapy per day, at least five times weekly. Patients in this setting are usually discharged to home or community settings after a shorter stay. Research into ASP in the acute rehabilitation setting is lacking. However, the themes and interventions described in this chapter are likely applicable, with modification, to acute rehabilitation facilities. Acute rehabilitation facilities housed within acute care hospitals may have more resources given their potential for proximity to specialized clinical pharmacists, infectious disease physicians, and on-site microbiology laboratories.

Compared to acute rehabilitation, sub-acute rehabilitation centers target patients with less intense therapy needs, who often participate in physical therapy for one to two hours daily. Direct contact with clinicians is less frequent compared to that in acute rehabilitation facilities. The length of stay is generally longer, and patients leave these facilities and return to the community or move into LTCFs. LTCF residents require regular skilled nursing assistance for activities of daily living. However, these residents do not require the ongoing, intense medical care that takes place in Long-Term Acute Care Hospitals such as prolonged ventilator weaning. Given the differences between care settings, we will point out areas that might be more applicable to rehabilitation centers rather than LTCFs and vice versa.

Existing Guidance

Multiple resources exist which can provide a general framework and aid in implementation of ASPs in long-term care settings including the Centers for Disease Control and Prevention's (CDC) Core Elements of Antibiotic Stewardship for Nursing Homes and the Agency for Healthcare Research and Quality (AHRQ)'s published guidance on ASP implementation specific to long-term care settings [12, 14]. Even more focused, the Society for Post-Acute and Long-Term Care Medicine created a template for policies tailored to LTCFs [15]. Direct links to these resources are listed at the end of this chapter for independent review. These tools provide useful guidance that can be tailored to a specific institution's resources and needs.

Infections in Long-Term Care Facilities and Rehabilitation Centers

Infections in LTCFs can be considered in two broad categories: infectious syndromes and problematic pathogens. Interventions that target these syndromes or infections are likely to be useful and lead to improvement in patient outcomes as well as possible cost savings [5, 9, 10, 16–18].

The three most commonly reported infectious syndromes in LTCFs are urinary tract infections (UTIs), skin and soft tissue infections (SSTIs), and respiratory tract infections (RTIs) [19–22]. This may stem from a high burden of residents with indwelling urinary catheters due to incontinence (UTIs), bedbound residents (SSTIs), and the close quarters inherent to many facilities (RTIs). Although acute rehabilitation facilities may have fewer bedbound patients, they may still see significant numbers of UTIs given a high number of patients with spinal cord injuries or strokes resulting in bladder incontinence and of RTIs given the close proximity of nursing and therapy staff with patients undergoing frequent therapeutic exercise along with communal dining and recreational areas.

Targeted pathogens that should be a focus of LTCF interventions include *Clostridioides difficile* associated colitis (CDI), methicillin-resistant *Staphylococcus aureus* (MRSA), and multi-drug resistant gram-negative bacteria (MDR GNB). Rising to importance in the last 20 years, MDR GNBs are of particular concern for two reasons. First, there are limited treatment options for infections with MDR GNBs and optimal treatment is often unclear [23]. Second, MDR GNB may harbor genes that produce resistance and can easily spread from one organism to another and from patient to patient. This transmissibility of resistance makes MDR GNB outbreaks within the close quarters of LTCFs and rehabilitation centers a significant threat to patients and of great importance from an infection prevention standpoint [24–26]. Prior antibiotic exposure is the most significant risk factor for colonization or infection with MDROs [24, 27–31].

Prior antibiotic exposure is also the greatest risk factor for *CDI* [32, 33]. CDI can cause significant distress to patients due to abdominal pain and high-volume diarrhea and is highly transmissible with mortality rates estimated between five and 30% in hospitalized patients [34, 35]. When exposed to antibiotics, normal microbial flora in the colon are depleted, and *Clostridioides difficile,* which occurs naturally in the digestive tract, multiplies leading to infection characterized by colitis and diarrhea. *Clostridioides difficile* produces spores that can pass from person to person even after an infected patient has left a room or bed [36, 37]. Cleaning practices that are effective for other infections may not kill these spores and specific cleaning products must therefore be used. Hand hygiene is less effective. Loose and frequent stools can contribute to decubitus ulcers in bedbound patients and cleaning these frequent stools places further burden on already thinly stretched facility staff. Furthermore, patients with significant diarrhea are reticent to participate in therapy, and their required isolation may reduce contact with therapists thereby delaying their rehabilitation.

Asymptomatic bacteriuria (ASB), although not an infection, brings together the frequently diagnosed "UTI" and the MDRO GNB threat. By definition, ASB is the presence of bacteria in urine without an associated infection. The Infectious Diseases Society of America (IDSA) has clear guidelines on avoiding treatment of ASB except in pregnant patients or those undergoing invasive urologic procedures [38]. ASB is nearly universally present in patients with chronic indwelling catheters and antibiotic therapy does not prevent recurrence or progression to infection [39–42]. Treating ASB has been shown in multiple studies to harm patients through lack of clinical benefit to the patient and resulting development of drug resistance [39, 41, 42]. The 2019 IDSA guidelines for ASB address ASB in functionally and cognitively impaired patients and strongly recommend against screening these patients for or treating ASB [38]. They also recommend clinicians evaluate for alternative causes of altered mental status or falls in these patients rather than attribute the change to bacteriuria [38]. Treating ASB is a common practice in some LTCFs and represents the proverbial "low-hanging fruit" for ASP intervention.

Now that we've defined common infections and syndromes present in LTCF and rehabilitation center patients, we can further discuss how to implement an ASP to combat these problems.

Steps for Antimicrobial Stewardship Program Implementation in Long-Term Care and Rehabilitation Facilities

1. Secure a Commitment
2. Create a Committee
3. Gather Data
4. Identify Areas for Improvement
5. Design and Implement Interventions
6. Measure and Re-assess
7. Educate, Educate, Educate

The intention of any ASP is to ensure antimicrobials are given to the right patients, in the right dosage, and for the right amount of time, which in turn will reduce adverse events associated with inappropriate usage.

To help LTCFs and rehabilitation centers achieve this goal of appropriate antimicrobial use through ASP, the CDC has created "Core Elements" for ASP programs that can be applied in both acute and long-term care settings. The core elements include: Leadership Commitment, Accountability, Pharmacy Expertise, Action, Tracking, Reporting and Education (Fig. 1.2).

We will emphasize when the steps described below clearly align with the CDC's Core Elements. Identifying these connections during your ASP implementation can help compliance tracking and monitoring for CMS or other regulators.

Step 1: Secure a Commitment

Ensuring the *commitment* from facility leadership in prioritizing ASP implementation is a crucial first step to creating a successful program [43] and is the first of the CDC's Core Elements of ASP. In long-term care facilities and rehabilitation centers, this would typically involve executive leadership, nursing leadership, and medical directors. Leadership can be encouraged to commit to ASP efforts through tailored pitches to highlight the positive outcomes such as (1) cost benefits for decreased antimicrobial usage and the costly sequelae of antimicrobial overuse, (2) meeting federally mandated ASP initiatives, and (3) the subsequent benefits from ASP

Fig. 1.2 CDC's core elements for antimicrobial stewardship programs

interventions in areas outside of intervention such as improved nursing–provider communication. Once leadership has committed to ASP, encourage them to show commitment via both public and internal displays. Outward examples of commitment can include signed letters displayed in public areas, communication to staff during regularly scheduled meetings to verbalize their investment in the ASP, and eventually celebrating improvements once ASP interventions are implemented. Internal examples of commitment can be allocating protected time to staff members involved in ASP as well as budgeting monetary support to the program. These steps not only ensure that those tasked with implementing stewardship will have the time and resources to devote to the task, they also create a culture of attention and investment in antimicrobial stewardship. When leadership makes a public display of commitment to the ASP, the success of the ASP becomes intertwined with that of leadership.

Step 2: Create a Committee

The next step involves structuring an appropriate Antimicrobial Stewardship team based on the size and resources of the facility. While making this team, the CDC's Core Elements of *accountability* and *drug expertise* should be considered. Typically, one or two individuals should be charged with *accountability* for the program's success. This individual should also serve as the stewardship champion—someone who promotes the ASP and drives interventions forward. This may be an invested committee member and often will be the medical director of the facility or designated physician. This structure is distinct from the more resourced acute care hospital, where ASP co-leadership between a pharmacist and physician is recommended. There are no clear guidelines for who should comprise additional team members, but appropriate choices include nurses who have leadership roles such as directors or charge nurses, medical directors (if not chair of the committee), consultant pharmacists, and infection preventionists. Infection preventionists must be given time and compensated for their effort outside of their concurrent responsibilities for infection prevention which, although related, includes separate tasks. Optional additional members could include but are not limited to frontline staff (nurses or nursing assistants), administrative staff, a board member or designee, and a representative from the resident and family council. Consideration should be given to include outside consultants such as prescribing clinicians already at the facility or those with specialized knowledge, including infectious diseases trained pharmacists or physicians. These specialists may already practice within the facility or could be community or hospital-based practitioners interested in providing expertise to the committee. These pharmacists or physicians with specialized infectious disease training are ideally suited to provide *drug expertise*. Surveys and studies of attitudes of LTCF staff and ASP programs have suggested that involvement of outside expertise may enhance the success of an ASP team and subsequent intervention [43, 44]. Facilities may employ outside infectious disease clinicians or pharmacists as consultants, or through direct employment on a part-time basis. Acute care rehabilitation centers housed within acute care hospitals can easily collaborate with the inpatient ASP team or experts in antibiotic stewardship. Given resource limitations, facilities may alternatively support existing staff to undergo additional training courses like Making a Difference in Infectious Diseases (MAD-ID) or attend conferences specializing in antimicrobial stewardship. Local and state health departments may be able to provide free or reduced cost educational resources on antimicrobial stewardship as well.

After a team is designated, initial review should ensure an understanding of the CDC's Core Elements and comprehension of the rationale for stewardship as not all team members may be familiar with ASP principles. Further roles that should be delegated to specific team members include the development and review of program tools and monitoring of program performance.

Step 3: Gather Data

Once team members are identified, the committee should perform a readiness assessment to help judge the feasibility of interventions. Leadership support, institutional stability, and level of resources including personnel, funds, and time allocation all inform which interventions are reasonable and most likely to effect sustained practice change. Next, data regarding current antimicrobial use and facility infections can be gathered. This can be done via reviewing existing medical records or examining pharmacy and prescribing records. Ideally an antibiogram would be created with help from each facility's laboratory partner. An antibiogram is a list of infectious organisms (usually bacteria) and their susceptibility patterns for the facility. An example is shown in Fig. 1.3 below. Smaller facilities may not have enough cultures in a given calendar year to create an antibiogram as the Clinical and Laboratory Standards Institute recommends at least 30 isolates from diagnostic specimens to create an appropriate one [45]. Note that specimens obtained for *screening* purposes should not be included in the antibiogram. Sites with only a few specimens can work to gather data over longer periods of time (years) to identify susceptibility patterns in their facility or combine regional data from multiple facilities [15, 46].

Areas to explore as potential sites of intervention can include (1) processes along the antibiotic decision-making pathway, (2) total and individual antibiotic usage, and (3) antibiotic outcomes [12]. Investigating process measures include reviewing the steps that lead to antibiotic prescription (Fig. 1.4). This would involve examining what assessment of the patient was made and by whom, what was prescribed and by whom, and whether the antibiotic selection was appropriate per clinical guidelines and institutional policies. To examine antibiotic usage, there are a number of potential metrics that can be collected. Selection of the appropriate metric or metrics will depend upon available resources (human and technologic). Options include percentage of antibiotics given intravenously (IV) vs via the mouth (or enteric tube), prevalence of residents receiving antibiotics on any given day (measured over time), number of new antibiotic starts per month, and days of therapy

Antibiogram of Skilled Nursing Home and Rehabilitation Center of Average City, USA

Organism	Antbiotic and % suseptbility				
Gram Positive Bacteria	**Cefazolin**	**Ceftriaxone**	**Vancomycin**	**Daptomycin**	**Linezolid**
Staphylococcus aureus	40%		100%	100%	95%
Streptococcus pneumoniae	56%	99%	82%		
Enterococcus faecalis	67%		99%	95%	100%
Enterococcus faecium	12%		70%	88%	100%

Gram Negative Bacteria	**Cefazolin**	**Ceftriaxone**	**Cefepime**	**Piperacillin-Tazobactam**	**Meropenem**
Escherichia coli	80%	90%	95%	96%	98%
Enterobacter cloacae	65%	73%	76%	82%	90%
Pseudomonas aeruginosa			74%	82%	99%
Proteus mirabilis	24%	56%	90%	92%	100%

Fig. 1.3 Sample antibiogram

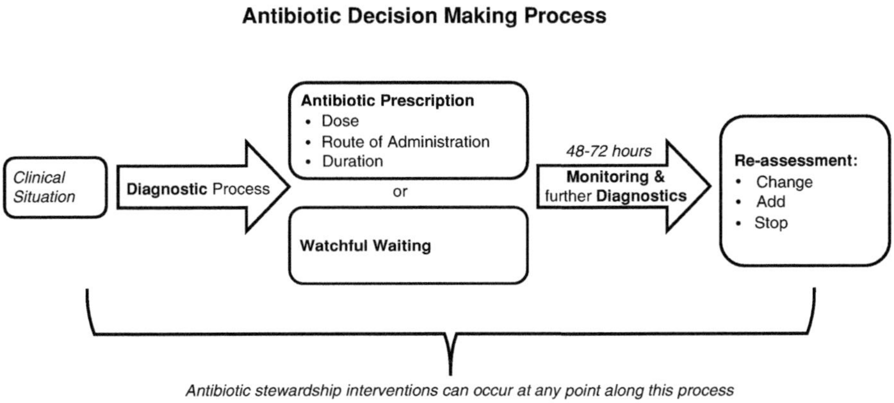

Fig. 1.4 Antibiotic decision-making process. Adapted from Sloane et al. NC Medical Journal 2016

(DOT) of antibiotics per 1000 person days. These measures can provide areas for improvement and can be tracked before and after changes are implemented. If your facility is within the CDC's National Healthcare Safety Network (NHSN), you may be able to partner with them to report into their Antibiotic Use Option which can help aggregate your antibiotic usage data and compare it to national benchmarks [47]. Similarly, participation in state programs, such as one run by the Massachusetts Department of Public Health which provides facilities with benchmarking of their monthly antibiotic start rates, can be a fruitful way to learn from experts, share ideas, and get feedback on antibiotic usage trends. Antibiotic outcome measurements include tracking adverse events such as CDI or MDR GNB infections and adverse drug events (including prescribing errors resulting in harm). Measuring these events and exploring your institution's data will provide a clearer picture of pre-existing success and opportunities for improvement that can become ASP interventions. If no one within the ASP is familiar with tracking and monitoring metrics, other committees or personnel within your institution may track and report quality metrics and could provide advice for the development and implementation of tracking processes. Outside groups familiar with quality improvement work, including consulting companies or administrators at a local or partnering healthcare facility, can provide guidance. The Institute for Healthcare Improvement also provides online training in quality improvement, which can be obtained at little to no cost (see resource list).

Step 4: Identify Areas for Improvement

Once the ASP team has identified areas of focus, including infections and resistance patterns, prescribing processes, and general antibiotic usage, they can select targets for improvement. ASP interventions can occur at any point along the decision-making process between a patient's clinical change to prescription re-assessment

(Fig. 1.4). Because each facility serves varying patient populations who require different services and levels of care, each facility's "optimal" use of antibiotics will vary. Furthermore, infection prevention efforts, which often work hand in hand with ASP initiatives, may be targeted to different problems. For instance, having a urinary catheter is well-established to be associated with UTIs. LTCF may want to work to reduce urinary catheter use and inappropriate treatment of asymptomatic bacteriuria, while an acute rehabilitation center may have fewer patients requiring urinary catheters making this intervention less meaningful. Conversely, acute rehabilitation facilities may have more patient congregation areas and higher numbers of prolonged close interactions between staff and patients, making such facilities more prone to outbreaks of respiratory viruses, which could be inappropriately treated with antibiotics. Acute rehabilitation facilities might therefore focus on reducing viral transmission via handwashing, mask wearing, isolation of sick patients, and concurrently reducing inappropriate antibiotic use for upper respiratory infections.

Other general areas of focus could include reducing the use of medications with increased risk of CDI (fluoroquinolones, clindamycin, cephalosporins, carbapenems), curbing the prescribing of antibiotics for indications where they are known to be contraindicated (e.g., viral respiratory infections and asymptomatic bacteriuria) and promoting the use of shorter courses based on national guidelines. Although commonly implemented in emergency departments and outpatient clinics with significant success [48–50], assessing and benchmarking the prescribing practices of individual clinicians may be difficult in LTCFs as on-call prescribers' orders may be attributed to a patient's primary provider or the facility's medical director. In rehabilitation centers where clinicians are more readily available on site, this assessment (coupled with audit and feedback) can be impactful. Committees should initially focus on one or two areas of optimization (see Fig. 1.5) that are relevant to their

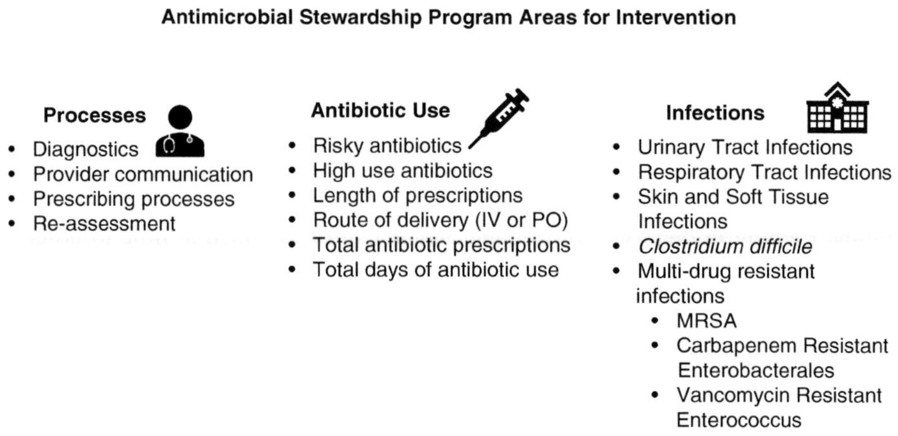

Fig. 1.5 Areas for intervention

facility (i.e., fluoroquinolone use and length of antibiotic therapy) in order to create meaningful interventions that can be completed in a reasonable amount of time to prove to leadership, staff, patients, and families the value of the ASP.

Step 5: Design and Implement Interventions

Once data gathering is established and areas for improvement are mapped out, interventions can be designed and implemented. The CDC recommends identifying small, easily measured outcomes that are rife for *action*. Below over-arching types of interventions are described in detail, including Diagnostic Stewardship and Management Algorithms, Antibiotic Utilization, and Antibiotic "time-outs." These general categories are just a few of the many areas of intervention. Interventions can be broad, such as improvement of all antibiotic ordering with standardized templates to include drug, indication, dose, frequency, and duration. Interventions can also be narrower in focus, such as syndrome-specific interventions targeting areas of concern such as UTI or CDI, or pharmacy focused, such as development of automated IV to oral antibiotic switch protocols. More in-depth examples of interventions are detailed below. Regardless of the plan chosen, detailed logistics should be mapped out. For example, if the Situation, Background, Assessment, and Request (SBAR) strategy for UTI is chosen (Fig. 1.6), details such as education for nurses, accessibility of the forms, communication with providers regarding the change, and data collection should all be planned prior to enactment. This provides for a more successful implementation and adequate data gathering, which will allow you to assess the effectiveness of your intervention and provide detailed reporting to stakeholders and regulatory bodies.

Suspected UTI SBAR

Situation

Hello. I am calling from (NAME OF FACILTY) about patient (NAME OF PATIENT) and a change in clinical status.

Vital Signs are BP ______/ ______ HR ______ Respiratory Rate______ Temperature ______

Background

Active diagnoses or other symptoms (especially bladder, kidney, genitourinary): ______

Resident has indwelling catheter ( ) YES ( ) NO

Resident is on dialysis ( ) YES ( ) NO

Resident is incontinent ( ) YES ( ) NO If yes, new or worsening? ( ) YES ( ) NO

Advance directives limiting treatment related to antibiotics or hospitalizations? ( ) YES ( ) NO
Specify: ______

Allergies? ( ) YES ( ) NO
Specify: ______

Suspected UTI SBAR

Assessment

Resident **WITH** indwelling catheter:
The criteria are met to initiate antibiotics if one of the below are selected

No	Yes	
( )	( )	Fever or 100°F (38°C) or repeated temperatures of 99°F (37°()
( )	( )	New back or flank pain
( )	( )	Acute pain
( )	( )	Rigors or shaking chills
( )	( )	New dramatic mental change
( )	( )	Hypotension (significant change from baseline BP or systolic BP <90mmHg)

Resident **WITHOUT** indwelling catheter:
The criteria are met if one of the three situations are met

No Yes

( ) ( ) 1. Acute dysuria

( ) ( ) 2. Fever of 100°F (38°C) **and** at least one new or worsening of the following:

( ) Urgency ( ) Suprapubic Pain
( ) Frequency ( ) Gross Hematuria
( ) Back/flank pain ( ) Urinary incontinence

( ) ( ) 3. No fever, but two or more of the following symptoms:

( ) Urgency ( ) Suprapubic Pain
( ) Frequency ( ) Gross Hematuria
( ) Back/flank pain ( ) Urinary incontinence

Nurses: Please check box to indicate whether criteria are met
( ) Protocols met. Resident may require urinalysis and culture and antibiotic
( ) Protocols NOT met. Resident does NOT need immediate antibiotic, but may need additional observation

Suspected UTI SBAR

Request

Order provided by clinician through ( ) Phone ( ) Fax ( ) In Person ( ) other ______
( ) Order urinalysis
( ) Order Urine Culture
( ) Encourage ______ ounces of liquid intake ______ times daily until urine is light yellow in color
( ) Record Fluid Intake
( ) Assess vital signs for ______ days, including temperature every ______ hours for ______ hours
( ) Notify Physician/NP/PA if symptoms worsen or are unresolved in ______ hours

( ) ***Initiate the following antibiotic:***
Antibiotic: ______ Dose: ______ Route: ______ Duration: ______
( ) No ( ) Yes Pharmacist to adjust for renal function

( ) OTHER ______

Physician/NP/PA Signature ______ Date/Time ______

Order Received by ______ Date/Time ______

Family/POA Notified by ______ Date/Time ______

Fig. 1.6 UTI SBAR form (Adapted from AHRQ)

Example Interventions

Diagnostic Stewardship and Management Algorithms

Diagnostic Stewardship simply means ensuring appropriate use of a diagnostic test. A commonly cited successful ASP diagnostic stewardship intervention has been to decrease inappropriate urine cultures in order to combat the inappropriate treatment of asymptomatic bacteriuria (ASB) [4, 51, 52]. Long-term care residents, especially but not only patients with chronic indwelling urinary catheters, are frequently colonized with bacteria that manifest as positive urine cultures in the absence of true infection [53]. Bacteria in the urine without clinical signs of infection (such as fever or hemodynamic instability not attributable to another cause, suprapubic pain or flank pain, or new hematuria) is called ASB [38, 39]. Giving patients with ASB antibiotics will not reduce the likelihood of subsequent infection and will lead to more MDRO colonization and potential for CDI [38, 41]. Reducing the number of urine cultures that are collected from residents without signs of true infection would be considered a type of diagnostic stewardship. Infection-specific interventions in diagnostic stewardship and management are described below. Each facility is encouraged to examine its own infection rates and determine if and how diagnostic stewardship can be applied.

Management algorithms provide clinicians with easy to follow steps to diagnose and treat infections. Ideally, they decrease prescribing variability and improve adherence to evidence-based guidelines. This principle has been applied to multiple types of infections. Management algorithms can include laminated pocket cards for clinicians that describe the appropriate empiric antibiotic regimen including dosing and length of treatment as well as guidance on physical interventions such as urinary catheter exchange for diagnosis of UTI. If your facility has an electronic medical record, this algorithm could simply be a built-in order-set. These can be created based on current evidence-based guidelines and/or with consultation from infectious disease experts (physician or pharmacist), and with attention to your local antibiogram.

Urinary Tract Infections

Loeb et al. reported their successful implementation of diagnostic and treatment algorithms for suspected UTI combined with interactive educational materials and personal meetings with staff across 24 nursing homes in Canada. They noted a reduction in number of courses of antibiotics prescribed for UTIs when compared to nursing homes who did not receive an intervention [54]. AHRQ has developed a tool employing diagnostic stewardship for UTIs named "UTI Situation, Background, Assessment, and Request (SBAR)" that is easily available and adaptable to LTCFs and rehabilitation centers. It consists of a form (Fig. 1.6) structured to facilitate communication between staff and prescribing clinicians to aid decision making about the need for additional testing and antibiotics. The assessment section directly addresses whether urine culture is appropriate and whether antimicrobial prescription or watchful waiting is preferred and serves as a modifiable example of diagnostic stewardship. This form, or a modified version, could be provided as laminated pocket cards for nurses to carry on the floor or have available at nursing stations. When a similar tool was used in 12 nursing homes in Texas, researchers noted a reduction in antibiotic prescriptions for ASB [55]. To implement this intervention and evaluate its efficacy, ASP teams could investigate its qualitative and quantitative effects. ASP members could interview providers and nurses regarding communication around the clinical status change of a patient with and without an indwelling urinary catheter before and after the intervention and provide qualitative data on the intervention. For quantitative evaluations, the number of urine cultures sent per time period could be calculated, and/or antibiotics prescribed for urinary tract infections, and/or total antibiotic prescribing days as UTIs are a common source of antibiotic prescriptions in LTCFs. Any intervention should be paired with education for nursing staff and providers to explain the intention of the intervention, signs, and symptoms associated with true UTI, and indications for testing and treatment. It is critical to ensure that the logistics of the intervention will not significantly interrupt workflow (i.e., where would the UTI SBAR form be located?), answer staff questions, and secure buy-in from clinicians by emphasizing the communication aspect of this intervention and need for close monitoring even if an antibiotic is not prescribed. It is important to highlight to all involved that this type of intervention should not only lead to decreased treatment of ASB and adverse outcomes associated with it but is also expected to improve overall communication between nurses and providers.

Respiratory Tract Infections

Lower respiratory tract infections (RTIs) and acute bronchitis are commonly caused by viral pathogens and antibiotics may not be indicated. Despite these well-established principles many clinicians prescribe antibiotics inappropriately for all RTIs. If your institution has a high rate of RTI diagnoses, a potential ASP intervention might be creating a diagnostic algorithm such as that presented in Fig. 1.7 to identify residents who would benefit from antibiotic administration and those whose symptoms could be monitored and antibiotics started only if they fail to improve. A complementary intervention might be in the form of a management algorithm to aid providers when deciding which antibiotic to prescribe. Recent IDSA guidelines provide recommendations on appropriate outpatient treatment of community acquired pneumonia (CAP), however many clinicians still employ levofloxacin as a first-line treatment for CAP despite the drug's black box warnings [57, 58]. A management algorithm can encourage physicians not to adhere to evidence-based guidelines and avoid dangerous medications. An example of a management algorithm for pneumonia is shown in Fig. 1.8.

Antibiotic Utilization

Antibiotic utilization is another potential ASP target. Interventions can target specific antibiotics that are costly to your center, overused or used inappropriately or target length of antibiotic treatment. Most infections require less than 7 days of treatment [57–61]. Providing antibiotics for a longer duration than is recommended by guidelines unnecessarily increases the risk of medication side effects, CDI, and antibiotic resistance. Interventions within this area may target healthcare associated infections such as CDI or MDROs as well as reducing overall costs to the facility.

Clinical Findings				Clinical Response	
Primary		Secondary		Primary and secondary both present	Either Primary or secondary absent
Afebrile	Other comorbidities	New productive cough +	Either of the following A. Respiratory rate > 25/min B. Delirium	Start antibiotics	Start active monitoring Re-evaluate next day (do not start antibiotics)
Afebrile	COPD pre-existing	Increased sputum purulence +	Either of the following A. Increased dyspnea B. Increased sputum volume		
Fever	>100°F or > 2.0°F above baseline or 2 x >99°F	>1 of the following symptoms (include at least 1 Respiratory Specific) **Respiratory Specific– at least one** -New or increase cough -New or increase sputum	**Respiratory Non-Specific** -Delirium -***Total WBC > 14,000**** -***Hypoxia (O2 sats < 90%)**** -Pleuritic chest pain -Respiratory rate > 25 breaths/min -Consolidation (on exam)	(Table 2 & 3)	(Table 2)
	High (>102F)	Respiratory Rate >25 **and/or New** Productive Cough		**Consider ED**	

* Secondary findings italicized may be documented/obtained by ordering these tests during active monitoring if resident initially does not meet criteria to start antibiotics
** Consider ordering chest X-ray and CBC with differential for febrile residents with cough and any of these criteria (HR >100, worsening mental status, or rigors). Antibiotics should not be used for up to 24 h after large-volume aspiration in those without COPD but with temp≤38.9°C (102 °F) and non-productive cough

TABLE 2. CLINICAL RESPONSE DETAILS

Start Antibiotics: **Primary & secondary findings** both present	**Start Active Monitoring:** Either Primary & secondary findings are **absent**
• **Test COVID PCR (+isolation) if still active** • **IF SEASONAL, √ RSV/INFLUENZA TESTING** • Choose antibiotic **(Table 3)** • Diagnostics only if o If community/hospital legionella problem–test urine Legionella o If not improving consider CXR, pulse oximetry, CBC	• **Do not start antibiotics** • **Consider COVID PCR test if still active monitoring** • If seasonal, consider ✓ RSV/influenza testing • If concerned or not improved ✓ **CBC+/–CXR** • Re-evaluate signs and symptoms every 24 hours. o If not improved, consider • Upper respiratory infection (URI) (if productive cough is not noted) • non-infectious causes of pulmonary infiltrate, other infection source.

* In patients where suspicion of Legionella is low, and the burden/risks of sputum or urine acquisition is high, testing not justified. Urinary catheterization, induced sputum production, or nasotracheal suctioning are only necessary in patients or residents when suspicion of Legionella is high and require proper consent.
2019 American Thoracic Society/IDSA guidelines: (1) suggest Legionella antigen testing in cases where a known outbreak is occurring or severe disease in hospitalized patients; (2) in adults with CAP< when influenza viruses are circulating in the community test for influenza with a rapid molecular assay.

Fig. 1.7 Management algorithm for lower respiratory infections created by Emory Antibiotic Stewardship in Long-term Care [56]

Table 3. Recommended agents and durations for Nursing Home Residents with Criteria to Start Antibiotics for Suspected Pneumonia				
Mild-Moderate				
	Scenario	Antibiotic to Start	Dosing and Duration (5 days)*	Considerations
1st line	Moderate illness, able to take oral medication**	Cefpodoxime	200 mg PO BID X 5 days	**Safe in mild PCN allergy** Q 24 hrs if CrCL <30; or post HD X 3 **Alternative cefuroxime 500 mg PO BID**
		Amoxicillin/clavulanate	500 mg/125 mg TID X 5 days	Equivalent to 875/125 mg BID
2nd line	Severe contraindications to 1st line	Levofloxacin	750 mg PO Q 24 X 5 days OR if CRCL<20 or ESRD, 750 x 1 dose, then 500 mg Q 48 hours X 2 doses	Pose a higher risk for C. difficile infection Caution with anti-arrhythmic medications and prolonged QTc Mild PCN allergy not an indication for quinolones
Special circumstances				
	Unable to take oral meds	Ceftriaxone IM/IV	1000 mg IM Q 24 hours X 5	Reserve IM for severe illness but reluctant to evaluate in ED/Hosptial
2nd line	Deterioration on oral agents, severe illness	Ceftriaxone IM/IV AND		
		Doxycycline OR Azithromycin	1000 mg BID X 5 days 500 mg X 3 days	No renal adjustment needed
	Risk for Pseudomonas***	Levofloxacin Monotherapy	As above	
	Risk MRSA***	Doxy + 1st line	1000 mg BID X 5 days + 1st line	Add doxy to first line agents.
	Risk of or suspect aspiration	Use either 1st line	Duration should be maintained at 5 days	Add anaerobic coverage only if abscess or empyema suspected: then amox/clavulanate, or ceftriaxone/metronixazole

* Duration is 5 days. Only extend to 7 if signs/symptoms not Improved at day 5 (i.e., still fever, use of supplemental O2, unstable vital signs).
we recommend B-lactam monotherapy. 2019 ATS/IDSA guidelines recommend combination therapy (b-lactam/macrolide) for outpatients with co-morbidities, but for nursing home residents without typical community-exposure to atypical organisms, min role for combination therapy, but not contraindicated. **Co-morbidities include chronic heart, lung, liver, renal disease; diabetes, alcoholism, malignancy, asplenia.
*** for MRSA = history of clinical culture with MRSA (sputum, wound, nasal) in past year; for Pseudomonas = history of + PSA in sputum in past year or bronchiectasis/FEV1<35%. If NOT transferring to ED to get new sputum culture, use levofloxacin (pseudomonas) or doxy + 1st line agent (for MRSA). **IF able** to order sputum microbiology, stop additional coverage if no Pseudomonas or MRSA recovered.

Fig. 1.8 Antibiotic management algorithm for pneumonia from Emory Antibiotic Stewardship in Long-term Care [56]

Clostridioides Difficile

As discussed previously, CDI occurs when the normal gut flora ("good" bacteria and yeast) is disrupted by antibiotics. Certain classes of antibiotics are known to encourage this dysbiosis including fluoroquinolones (such as ciprofloxacin and levofloxacin), later generation cephalosporins (such as ceftriaxone and cefepime), and carbapenems (such as meropenem) [62, 63]. Restricting these drugs may lower

CDI rates [64–67]. Fluoroquinolones have significant side effects including altered mental status, abnormal blood glucose, abnormal heart conduction (QTc prolongation) resulting in life-threatening arrhythmias, and aortic abnormalities, which have resulted in black box warnings from the Food and Drug Administration [68]. Despite these warnings, fluoroquinolones are still frequently used to empirically treat UTIs. Restricting their use by requiring a pharmacist approval or documented infection susceptible to these drugs (i.e., not allowing empiric utilization) may lower the rate of usage in your institution. Undoubtedly when usually available drugs become less available or unavailable, prescribers may become angry. In order to improve buy-in to the change and decrease resentment, we recommend education, discussion, and sometimes personal feedback *prior to* implementing restrictions. This can include town-hall style meetings, coffee talks, institution wide emails, or highly visible posters to explain the change and rationale behind it. If data are available, teams can use the audit and feedback technique to identify high-use prescribers to demonstrate how their practice differs from that of their peers and the standard of care and inform them of alternative prescribing options. These interventions are both effective and create sustained changes in practice [69, 70]. Ensuring providers know and have easy access to alternative medications may improve uptake of the intervention; therefore providing resources (such as a management algorithm as described previously) that give alternative antibiotic options along with dosing and length of treatment recommendations can improve buy-in and adherence.

Antibiotic Time-Out

Another well-described intervention includes an antibiotic "time-out" whereby staff reassesses a patient's need for antibiotics after two to three days of usage. Often due to lack of resources such as point-of-care laboratory testing or X-ray machines in a LTCF or rehabilitation center, antibiotics are started empirically but then remain prescribed even if upon further evaluation the suspicion for infection is low. To combat this, a process could be implemented whereby staff members review the need for antibiotics at three and seven days to see if they remain necessary. This would also provide an opportunity for direct feedback and education to providers regarding appropriate selection of antibiotic for a given syndrome and length of use. This intervention has been shown to be effective in reducing rates of targeted antibiotics and length of treatment in the acute care setting and can be driven by nurses, pharmacists, or infectious disease physicians [71, 72]. Multiple time-out tools are available for free on the internet.

Local infectious disease experts can be of great use for antibiotic utilization planning and antibiotic time-outs. Tedeschi et al. described an intervention of systematic bedside infectious disease consultation for all patients at an acute rehabilitation facility who were prescribed an antibiotic. This was combined with revisions of protocols along with education relating to antibiotic use. Their multi-pronged approach successfully decreased CDI, prevalence of highly drug-resistant bacteria, and overall antibiotic use [73]. Given the explosion of telemedicine, these experts can be engaged in-person or electronically to provide face-to-face assessments no matter how remote your facility is from these specialists.

All the previously described interventions should be tailored for your institution's needs, culture, and antibiotic prescribing habits. Common themes include working to ensure staff buy-in through education and involvement as well as clear leadership investment. Involving experts in infectious disease can ensure interventions adhere to evidence-based guidelines and may lend authority to interventions. As you can see, many interventions can be combined to try to enact sustained change and combat inappropriate antibiotic use at multiple points from diagnosis, to prescription, to length of treatment. In fact, bundling these interventions may provide the highest yield; Nace and colleagues implemented multi-facility interventions including webinars, educational tools, guideline concordant order sets, as well as monthly coaching sessions, and regular feedback reports. These interventions successfully decreased inappropriate antibiotic use and CDI without negatively impacting hospitalization or death [51]. Deciding what works for your facility is one of the ASP's most important jobs.

Step 6: Measure and Re-Assess

After a particular tool or intervention is implemented, regular meetings should be held to review data, troubleshoot performance, and provide feedback to institutional stakeholders. These meetings will be opportunities for the ASP to engage in the Core Elements of *tracking* and *reporting*. While data tracking will be an important part of each intervention, additional data can be reviewed outside the focus of the intervention such as number of infections in the categories mentioned above (UTI, RTI, CDI), use of high-risk drugs (fluoroquinolones, cephalosporins, carbapenems), and length of antibiotic prescriptions. These broad metrics may be indirectly impacted by targeted interventions and thus should be monitored for both positive and negative changes. Other data can be collected including provider and nursing staff attitudes toward ASP interventions, rates of hospital transfer, and cost to the facility related to antibiotic usage. The data collected as part of interventions and other utilization metrics should be reported back to nursing staff, clinical providers, and hospital leadership to further reinforce the importance of the antibiotic stewardship committee's work. It can also provide motivation to facility staff and serves to comply with CMS requirements regarding reporting [15].

Studies have shown that as time passes since an intervention is implemented the effect of that intervention decreases [54]. It is therefore important to continue to track, discuss, report, and re-affirm effective interventions to not lose initial gains. Analysis of outcomes will allow the committee to evaluate successes and areas for growth. These should in turn guide modification to current interventions or guide new projects.

Step 7: Educate, Educate, Educate

A continual step in ASP implementation is targeted *education* for nursing staff, clinical providers, and residents and families. Education can come in many forms and should continue throughout an intervention period. Less resource intense efforts can include mounting flyers, providing laminated antibiotic guides, or regular electronic communications. Interactive didactics are more resource intense but have greater evidence for improvements in antibiotic utilization. Residents and families should also be incorporated in education efforts as they can be fierce advocates for themselves and their loved ones [12]. This could be done with flyers in patient rooms or pamphlets discussing antibiotic use and resistance provided to residents and family members upon admission. By obtaining patient and family buy-in to stewardship efforts, programs can reduce expectations to prescribe antibiotics for residents when they may not be indicated and reinforce sound stewardship practices to bedside clinical staff. Change from educational interventions may be more difficult to sustain, therefore it is important to reinforce education by rewarding practice changes and providing reminders.

Conclusion

Antibiotic overutilization and the resultant harms are of great concern to management, staff, as well as residents and their families at LTCFs and rehabilitation centers. ASPs can help these facilities manage their antibiotic usage and empower staff to provide high-quality, evidence-based care. Furthermore, CMS requires all associated facilities to create an ASP to maintain accreditation for reimbursement. We hope that this chapter has provided the required background information regarding the importance of ASPs as well as steps to create ASPs and possible interventions that are applicable to LTCFs and rehabilitation centers.

Resources

Center for Disease Control and Prevention's Core Elements of Antibiotic Stewardship in Nursing Homes. https://www.cdc.gov/longtermcare/prevention/antibiotic-stewardship.html#anchor_1557415584.

Agency for Healthcare Research and Quality's Nursing Home Antibiotic Stewardship Guide. https://www.ahrq.gov/nhguide/index.html.

Institute for Healthcare Improvement Open School. http://www.ihi.org/education/IHIOpenSchool/Pages/default.aspx.

Infection Advisory Committee for AMDA—The Society of Post-Acute and Long-Term Care Medicine. https://spice.unc.edu/wp-content/uploads/2017/10/Jump-et-al-Antibiotic-Stewardship-Template-_JAMDA-2017.pdf.

Massachusetts Long-Term Care Facility Antibiotic Stewardship Resource Hub. https://mainfectioncontrol.populationhealthexchange.org/other-topics/ltcf-as/.

References

1. Strausbaugh LJ, Joseph CL. The burden of infection in long-term care. Infect Control Hosp Epidemiol. 2000;21(10):674–9. https://doi.org/10.1086/501712.
2. Nicolle LE, Bentley DW, Garibaldi R, Neuhaus EG, Smith PW. Antimicrobial use in long-term-care facilities. SHEA Long-term-care committee. Infect Control Hosp Epidemiol. 2000;21(8):537–45. https://doi.org/10.1086/501798.
3. Daneman N, Bronskill SE, Gruneir A, et al. Variability in antibiotic use across nursing homes and the risk of antibiotic-related adverse outcomes for individual residents. JAMA Intern Med. 2015;175(8):1331. https://doi.org/10.1001/jamainternmed.2015.2770.
4. Lewis, Steven, Agarwal, Nitika, Afolarin, Helen, Isenberg, Susan, Strong, Sally. Antibiotic stewardship in an inpatient rehabilitation hospital. 2012. https://idsa.confex.com/idsa/2012/webprogram/Paper37198.html.
5. Talkington K, David H, Deepa R. A path to better antibiotic stewardship in inpatient settings. Pew Chartitable Trusts; 2016. p. 45.
6. Cooper SJ. Overview of CMS rule changes to promote; 2020. https://www.hhs.gov/sites/default/files/1.3-cooper-cop-changes-for-abx-steward-508.pdf. Accessed 6 Aug 2021.
7. Fu CJ, Mantell E, Stone PW, Agarwal M. Characteristics of nursing homes with comprehensive antibiotic stewardship programs: results of a national survey. Am J Infect Control. 2020;48(1):13–8. https://doi.org/10.1016/j.ajic.2019.07.015.
8. Nicolle LE, Strausbaugh LJ, Garibaldi RA. Infections and antibiotic resistance in nursing homes. Clin Microbiol Rev. 1996;9(1):1–17.
9. Sokumbi O, Wetter DA. Clinical features, diagnosis, and treatment of erythema multiforme: a review for the practicing dermatologist. Int J Dermatol. 2012;51(8):889–902. https://doi.org/10.1111/j.1365-4632.2011.05348.x.
10. Slayton RB, Scott RD, Baggs J, Lessa FC, McDonald LC, Jernigan JA. The cost–benefit of federal investment in preventing clostridium difficile infections through the use of a multifaceted infection control and antimicrobial stewardship program. Infect Control Hosp Epidemiol. 2015;36(6):681–7. https://doi.org/10.1017/ice.2015.43.
11. Smith R, Coast J. The true cost of antimicrobial resistance. BMJ. 2013;346:f1493. https://doi.org/10.1136/bmj.f1493.
12. Core Elements of Antibiotic Stewardship for Nursing Homes I LTCF I CDC; 2021. https://www.cdc.gov/longtermcare/prevention/antibiotic-stewardship.html. Accessed 13 July 2021.
13. Daneman N, Gruneir A, Bronskill SE, et al. Prolonged antibiotic treatment in long-term care: role of the prescriber. JAMA Intern Med. 2013;173(8):673–82.
14. Agency for Healthcare Research and Quality. Nursing home antimicrobial stewardship guide. Agency for Healthcare Research and Quality. https://www.ahrq.gov/nhguide/index.html. Accessed 13 July 2021.
15. Jump RLP, Gaur S, Katz MJ, et al. Template for an antibiotic stewardship policy for post-acute and long-term care settings. J Am Med Dir Assoc. 2017;18(11):913–20. https://doi.org/10.1016/j.jamda.2017.07.018.
16. Borde JP, Nussbaum S, Hauser S, et al. Implementing an intensified antibiotic stewardship programme targeting daptomycin use in orthopaedic surgery: a cost–benefit analysis from the hospital perspective. Infection. 2016;44(3):301–7. https://doi.org/10.1007/s15010-015-0854-y.
17. Kravitz GR. Making the business case for ASP: taking it to the C-suite. Presented at: SHEA. https://www.shea-online.org/images/priority-topics/Business_Case_for_ASP.pdf. Accessed 16 Sept 2021.
18. Sick AC, Lehmann CU, Tamma PD, Lee CKK, Agwu AL. Sustained savings from a longitudinal cost analysis of an internet-based preapproval antimicrobial stewardship program. Infect Control Hosp Epidemiol. 2013;34(6):573–80. https://doi.org/10.1086/670625.
19. Chen H, Chiu AP, Lam PS, et al. Prevalence of infections in residential care homes for the elderly in Hong Kong. Hong Kong Med J. 2008;14(6):444–50.

20. Cotter M, Donlon S, Roche F, Byrne H, Fitzpatrick F. Healthcare-associated infection in Irish long-term care facilities: results from the First National Prevalence Study. J Hosp Infect. 2012;80(3):212–6. https://doi.org/10.1016/j.jhin.2011.12.010.
21. Roberts C, Roberts J, Roberts R. Survey of healthcare-associated infection rates in a nursing home resident population. J Infect Prev. 2010;11(3):82–6. https://doi.org/10.1177/1757177410364867.
22. Juthani-Mehta M, Quagliarello VJ. Infectious diseases in the nursing home setting: challenges and opportunities for clinical investigation. Clin Infect Dis. 2010;51(8):931–6. https://doi.org/10.1086/656411.
23. Tamma PD, Aitken SL, Bonomo RA, Mathers AJ, van Duin D, Clancy CJ. Infectious Diseases Society of America antimicrobial resistant treatment guidance: gram-negative bacterial infections. Clin Infect Dis. 2021;72(7):e169–83. https://doi.org/10.1093/cid/ciaa1478.
24. O'Fallon E, Schreiber R, Kandel R, D'Agata EMC. Multidrug-resistant gram-negative bacteria at a long-term care facility: assessment of residents, healthcare workers, and inanimate surfaces. Infect Control Hosp Epidemiol. 2009;30(12):1172–9. https://doi.org/10.1086/648453.
25. Logan LK, Weinstein RA. The epidemiology of carbapenem-resistant enterobacteriaceae: the impact and evolution of a global menace. J Infect Dis. 2017;215(suppl_1):S28–36. https://doi.org/10.1093/infdis/jiw282.
26. D'Agata EMC, Habtemariam D, Mitchell S. Multidrug-resistant gram-negative bacteria: inter- and intra-dissemination among residents with advanced dementia. Infect Control Hosp Epidemiol. 2015;36(8):930–5. https://doi.org/10.1017/ice.2015.97.
27. Benenson S, Cohen MJ, Block C, et al. Vancomycin-resistant enterococci in long-term care facilities. Infect Control Hosp Epidemiol. 2009;30(8):786–9. https://doi.org/10.1086/598345.
28. Mendelson G, Hait V, Ben-Israel J, Gronich D, Granot E, Raz R. Prevalence and risk factors of extended-spectrum beta-lactamase-producing Escherichia coli and Klebsiella pneumoniae in an Israeli long-term care facility. Eur J Clin Microbiol Infect Dis. 2005;24(1):17–22. https://doi.org/10.1007/s10096-004-1264-8.
29. Cohen AE, Lautenbach E, Morales KH, Linkin DR. Fluoroquinolone-resistant Escherichia coli in the long-term care setting. Am J Med. 2006;119(11):958–63. https://doi.org/10.1016/j.amjmed.2006.05.030.
30. Loeb MB, Craven S, McGeer AJ, et al. Risk factors for resistance to antimicrobial agents among nursing home residents. Am J Epidemiol. 2003;157(1):40–7. https://doi.org/10.1093/aje/kwf173.
31. Elizaga ML, Weinstein RA, Hayden MK. Patients in long-term care facilities: a reservoir for vancomycin-resistant enterococci. Clin Infect Dis. 2002;34(4):441–6. https://doi.org/10.1086/338461.
32. Czepiel J, Dróżdż M, Pituch H, et al. Clostridium difficile infection: review. Eur J Clin Microbiol Infect Dis. 2019;38(7):1211–21. https://doi.org/10.1007/s10096-019-03539-6.
33. Center for Disease Control. Clostridioides difficile. https://www.cdc.gov/cdiff/pdf/Cdiff-Factsheet-P.pdf. Accessed 8 Oct 2021.
34. Chu YM, Lee CL, Chen HY, Hung CS. Predictors of mortality in patients with Clostridium difficile infection. Adv Digestive Med. 2020;7(2):77–82. https://doi.org/10.1002/aid2.13159.
35. Appaneal HJ, Caffrey AR, Beganovic M, Avramovic S, LaPlante KL. Predictors of mortality among a national cohort of veterans with recurrent clostridium difficile infection. Open Forum Infect Dis. 2018;5(8):ofy175. https://doi.org/10.1093/ofid/ofy175.
36. Freedberg DE, Salmasian H, Cohen B, Abrams JA, Larson EL. Receipt of antibiotics in hospitalized patients and risk for clostridium difficile infection in subsequent patients who occupy the same bed. JAMA Intern Med. 2016;176(12):1801–8. https://doi.org/10.1001/jamainternmed.2016.6193.
37. Tarrant J, Jenkins RO, Laird KT. From ward to washer: the survival of Clostridium difficile spores on hospital bed sheets through a commercial UK NHS healthcare laundry process. Infect Control Hosp Epidemiol. 2018;39(12):1406–11. https://doi.org/10.1017/ice.2018.255.
38. Nicolle LE, Gupta K, Bradley SF, et al. Clinical practice guideline for the management of asymptomatic bacteriuria: 2019 update by the Infectious Diseases Society of America. Clin Infect Dis. 2019;68(10):e83–e110. https://doi.org/10.1093/cid/ciy1121.

39. Cortes-Penfield NW, Trautner BW, Jump R. Urinary tract infection and asymptomatic bacteriuria in older adults. Infect Dis Clin N Am. 2017;31(4):673–88. https://doi.org/10.1016/j.idc.2017.07.002.
40. Nicolle LE, Mayhew WJ, Bryan L. Prospective randomized comparison of therapy and no therapy for asymptomatic bacteriuria in institutionalized elderly women. Am J Med. 1987;83(1):27–33. https://doi.org/10.1016/0002-9343(87)90493-1.
41. Abrutyn E, Berlin J, Mossey J, Pitsakis P, Levison M, Kaye D. Does treatment of asymptomatic bacteriuria in older ambulatory women reduce subsequent symptoms of urinary tract infection? J Am Geriatr Soc. 1996;44(3):293–5. https://doi.org/10.1111/j.1532-5415.1996.tb00917.x.
42. Das R, Towle V, Van Ness PH, Juthani-Mehta M. Adverse outcomes in nursing home residents with increased episodes of observed bacteriuria. Infect Control Hosp Epidemiol. 2011;32(1):84–6. https://doi.org/10.1086/657664.
43. Seshadri S, Felsen CB, Sellers CR, Dumyati GK. "There is no one to pick up the pieces": sustainability of antibiotic stewardship programs in nursing homes. Infect Control Hosp Epidemiol. 2021;42(4):440–7. https://doi.org/10.1017/ice.2020.1217.
44. Felsen CB, Dodds Ashley ES, Barney GR, et al. Reducing fluoroquinolone use and Clostridioides difficile infections in community nursing homes through hospital–nursing home collaboration. J Am Med Dir Assoc. 2020;21(1):55–61.e2. https://doi.org/10.1016/j.jamda.2019.11.010.
45. M100Ed31 | Performance standards for antimicrobial susceptibility testing. 31st ed. Clinical & Laboratory Standards Institute. https://clsi.org/standards/products/microbiology/documents/m100/. Accessed 5 Aug 2021.
46. Fridkin SK, Pack J, Licitra G, et al. Creating reasonable antibiograms for antibiotic stewardship programs in nursing homes: analysis of 260 facilities in a large geographic region, 2016-2017. Infect Control Hosp Epidemiol. 2019;40(8):839–46. https://doi.org/10.1017/ice.2019.112.
47. The National Healthcare Safety Network. Long term care facility component tracking infections in long-term care facilities; 2020. https://www.cdc.gov/nhsn/pdfs/validation/2020/2020-ltcf-manual-508.pdf. Accessed 16 Sept 2021.
48. Buehrle DJ, Shively NR, Wagener MM, Clancy CJ, Decker BK. Sustained reductions in overall and unnecessary antibiotic prescribing at primary care clinics in a veterans affairs healthcare system following a multifaceted stewardship intervention. Clin Infect Dis. 2020;71(8):e316–22. https://doi.org/10.1093/cid/ciz1180.
49. Meeker D, Linder JA, Fox CR, et al. Effect of behavioral interventions on inappropriate antibiotic prescribing among primary care practices: a randomized clinical trial. JAMA. 2016;315(6):562–70. https://doi.org/10.1001/jama.2016.0275.
50. Hallsworth M, Chadborn T, Sallis A, et al. Provision of social norm feedback to high prescribers of antibiotics in general practice: a pragmatic national randomised controlled trial. Lancet. 2016;387(10029):1743–52. https://doi.org/10.1016/S0140-6736(16)00215-4.
51. Nace DA, Hanlon JT, Crnich CJ, et al. A multifaceted antimicrobial stewardship program for the treatment of uncomplicated cystitis in nursing home residents. JAMA Intern Med. 2020;180(7):944. https://doi.org/10.1001/jamainternmed.2020.1256.
52. Peoples K. Impact of successful implementation of an antimicrobial stewardship program in an inpatient rehabilitation setting. Am J Infect Control. 2019;47(6):S16. https://doi.org/10.1016/j.ajic.2019.04.015.
53. Phillips CD, Adepoju O, Stone N, et al. Asymptomatic bacteriuria, antibiotic use, and suspected urinary tract infections in four nursing homes. BMC Geriatr. 2012;12:73. https://doi.org/10.1186/1471-2318-12-73.
54. Loeb M, Brazil K, Lohfeld L, et al. Effect of a multifaceted intervention on number of antimicrobial prescriptions for suspected urinary tract infections in residents of nursing homes: cluster randomised controlled trial. BMJ. 2005;331(7518):669. https://doi.org/10.1136/bmj.38602.586343.55.
55. McMaughan DK, Nwaiwu O, Zhao H, et al. Impact of a decision-making aid for suspected urinary tract infections on antibiotic overuse in nursing homes. BMC Geriatr. 2016;16:81. https://doi.org/10.1186/s12877-016-0255-9.

56. Emory Antibiotic Stewardship in Long-term Care. Active monitoring guide. https://med.emory.edu/departments/medicine/divisions/infectious-diseases/easil/education.html. Accessed 7 Oct 2021.
57. Metlay JP, Waterer GW, Long AC, et al. Diagnosis and treatment of adults with community-acquired pneumonia. An official clinical practice guideline of the American Thoracic Society and Infectious Diseases Society of America. Am J Respir Crit Care Med. 2019;200(7):e45–67. https://doi.org/10.1164/rccm.201908-1581ST.
58. Vaughn VM, Gandhi TN, Chopra V, et al. Antibiotic overuse after hospital discharge: a multi-hospital cohort study. Clin Infect Dis. 2020;73:ciaa1372. https://doi.org/10.1093/cid/ciaa1372.
59. Gupta N, Limbago BM, Patel JB, Kallen AJ. Carbapenem-resistant enterobacteriaceae: epidemiology and prevention. Clin Infect Dis. 2011;53(1):60–7. https://doi.org/10.1093/cid/cir202.
60. Yahav D, Franceschini E, Koppel F, et al. Seven versus 14 days of antibiotic therapy for uncomplicated gram-negative bacteremia: a noninferiority randomized controlled trial. Clin Infect Dis. 2019;69(7):1091–8. https://doi.org/10.1093/cid/ciy1054.
61. Kalil AC, Metersky ML, Klompas M, et al. Management of adults with hospital-acquired and ventilator-associated pneumonia: 2016 clinical practice guidelines by the Infectious Diseases Society of America and the American Thoracic Society. Clin Infect Dis. 2016;63(5):e61–e111. https://doi.org/10.1093/cid/ciw353.
62. Owens RC Jr, Donskey CJ, Gaynes RP, Loo VG, Muto CA. Antimicrobial-associated risk factors for clostridium difficile infection. Clin Infect Dis. 2008;46(Supplement_1):S19–31. https://doi.org/10.1086/521859.
63. Brown KA, Khanafer N, Daneman N, Fisman DN. Meta-analysis of antibiotics and the risk of community-associated clostridium difficile infection. Antimicrob Agents Chemother. 2013;57(5):2326–32. https://doi.org/10.1128/AAC.02176-12.
64. Feazel LM, Malhotra A, Perencevich EN, Kaboli P, Diekema DJ, Schweizer ML. Effect of antibiotic stewardship programmes on Clostridium difficile incidence: a systematic review and meta-analysis. J Antimicrob Chemother. 2014;69(7):1748–54. https://doi.org/10.1093/jac/dku046.
65. Climo MW, Israel DS, Wong ES, Williams D, Coudron P, Markowitz SM. Hospital-wide restriction of clindamycin: effect on the incidence of Clostridium difficile-associated diarrhea and cost. Ann Intern Med. 1998;128(12 Pt 1):989–95. https://doi.org/10.7326/0003-4819-128-12_part_1-199806150-00005.
66. Valiquette L, Cossette B, Garant MP, Diab H, Pépin J. Impact of a reduction in the use of high-risk antibiotics on the course of an epidemic of Clostridium difficile-associated disease caused by the hypervirulent NAP1/027 strain. Clin Infect Dis. 2007;45(Suppl 2):S112–21. https://doi.org/10.1086/519258.
67. Dancer SJ, Kirkpatrick P, Corcoran DS, Christison F, Farmer D, Robertson C. Approaching zero: temporal effects of a restrictive antibiotic policy on hospital-acquired Clostridium difficile, extended-spectrum β-lactamase-producing coliforms and meticillin-resistant Staphylococcus aureus. Int J Antimicrob Agents. 2013;41(2):137–42.
68. Commissioner O of the. FDA updates warnings for fluoroquinolone antibiotics on risks of mental health and low blood sugar adverse reactions. FDA; 2020. https://www.fda.gov/news-events/press-announcements/fda-updates-warnings-fluoroquinolone-antibiotics-risks-mental-health-and-low-blood-sugar-adverse. Accessed 6 Aug 2021.
69. Schwartz DN, Abiad H, DeMarais PL, et al. An educational intervention to improve antimicrobial use in a hospital-based long-term care facility. J Am Geriatr Soc. 2007;55(8):1236–42. https://doi.org/10.1111/j.1532-5415.2007.01251.x.
70. Monette J, Miller MA, Monette M, et al. Effect of an educational intervention on optimizing antibiotic prescribing in long-term care facilities. J Am Geriatr Soc. 2007;55(8):1231–5. https://doi.org/10.1111/j.1532-5415.2007.01250.x.

71. Schooneveld TCV, Rupp ME, Cavaleiri RJ, Lyden E, Rolek K. Cluster randomized trial of an antibiotic time-out led by a team-based pharmacist. Infect Control Hosp Epidemiol. 2020;41(11):1266–71. https://doi.org/10.1017/ice.2020.347.
72. Vasina L, Dehner M, Wong A, et al. The impact of a pharmacist driven 48-hour antibiotic time out during multi-disciplinary rounds on antibiotic utilization in a community non-teaching hospital. Open Forum Infect Dis. 2017;4(Suppl 1):S272–3. https://doi.org/10.1093/ofid/ofx163.605.
73. Tedeschi S, Trapani F, Giannella M, et al. An antimicrobial stewardship program based on systematic infectious disease consultation in a rehabilitation facility. Infect Control Hosp Epidemiol. 2017;38(1):76–82. https://doi.org/10.1017/ice.2016.233.

Antimicrobial Stewardship in the Outpatient Setting

2

Monika Zmarlicka, Jacinta Chin, and Gabriela Andujar Vazquez

Antimicrobial resistance continues to grow as concern, threatening the utility of currently available life-saving antibiotics [1]. It is estimated that antibiotic-resistant pathogens are responsible for over two million infections within the USA alone, and over 20,000 deaths annually. The Centers for Disease Control and Prevention (CDC) has identified improved antibiotic prescribing practices via antimicrobial stewardship (AS) to be an essential piece in the fight against antimicrobial resistance [1]. It is estimated that approximately 60% of antibiotic prescriptions are generated in the outpatient setting [2]. Several studies have examined appropriateness of antibiotic prescribing in the outpatient settings and have found that a sizeable proportion of these prescriptions was unnecessary—one study estimated that 46% of antibiotic prescriptions in the studied population were not needed [3]. Thus, it is evident from the literature there is ample room for AS to intervene to help improve antibiotic prescribing practices in the outpatient setting.

M. Zmarlicka (✉)
Department of Pharmacy, Valleywise Health, Phoenix, AZ, USA
e-mail: Monika.zmarlicka@valleywisehealth.org

J. Chin
Department of Pharmacy, Novant Health, Charlotte, NC, USA
e-mail: jacintachin@novanthealth.org

G. A. Vazquez
Division of Geographic Medicine and Infectious Diseases at Tufts Medical Center, Boston, MA, USA

Tufts University School of Medicine, Boston, MA, USA
e-mail: gandujarvazquez@tuftsmedicalcenter.org

S. Doron, M. Campion (eds.), *Antimicrobial Stewardship in Non-Traditional Settings*, https://doi.org/10.1007/978-3-031-21710-4_2

The Essentials of Stewardship in the Outpatient Setting

A call for developing AS programs throughout the healthcare continuum has become a national priority in order to combat antimicrobial resistance [4]. Guidance on how to implement an outpatient AS program is minimal and data on its effectiveness and cost is scarce. In 2016, the CDC released the core elements of Outpatient AS for facilities to use as a reference point into starting an AS program [5]. Consensus has been reached that a multidisciplinary, multifaceted approach is most effective to improve antibiotic use [6].

Establishing an AS team is the first step to starting a program [7]. Ideally, team members should include pharmacists and physicians, preferably with infectious diseases training, and administrative, and nursing leadership if able. Identifying key stakeholders for an AS program to be successful is critical for sustainability and feasibility. These stakeholders can be identified for a specific role in a task force that pertains to their job description or as part of the AS team in general. Once team members have been identified and established, different elements to implementing an AS program can be assessed [5]. Commitment is one of the four CDC core elements of outpatient AS. Hospital leadership support has proven to play a crucial role in the success of antimicrobial stewardship programs. A written statement of commitment to AS by administration and clinical leadership has shown to have significant impact in achieving a robust, effective, and comprehensive AS program [8].

A comprehensive educational program is a key component of an AS program [7]. This should include not only prescribers and pharmacists, but also nurses, other health professionals, administrative staff, patients, and families. Over the years, it has been recognized that nursing staff are critical to AS efforts. They can play an essential role in several activities such as optimizing testing when needed, assuring optimal specimen collection for culture and making sure they were performed before antibiotic initiation, prompting discussions of antibiotic treatment, indication and duration, as well as improving evaluation of documentation of antibiotic allergies, among other things. Since antimicrobials affect every individual, whether patient, clinical or non-clinical staff, stewardship of these should be in everyone's interest [9–11].

Education could be approached in different ways in order to target as much audience as possible. Successes have been achieved with different strategies such as displaying posters at the waiting room or through activities like conference presentations, webinars, workshops, providing written guidelines, newsletters, email alerts, etc. [12–16]. Another educational strategy that has proven to be successful is communication training for prescribers. Giving prescribers the communication skills to navigate difficult conversations with their patients and families can decrease unnecessary antibiotic prescribing by reducing the perceived expectation from patient and families to receive an antibiotic [17]. Strategies like allowing for delayed prescribing with a call after an encounter in order to assess progression might be helpful in avoiding unnecessary antibiotic use. Reassurance to both provider and patient can improve patient safety and satisfaction [18, 19]. A caveat to education in antibiotic stewardship is that it should be continuous and always combined with other interventions since it alone is not as effective as it tends to diminish over time without reinforcement. There are multiple educational tools and resources available online through the CDC and local and state department of public health that can be used and adapted to each particular need.

In the outpatient setting the majority of the antibiotic prescribing happens empirically with the majority being broad spectrum [20]. One way to help the choice of empiric antibiotics is through the use of antibiograms. There are challenges to getting an accurate and useful antibiogram in the outpatient setting since most do not have enough isolates or capacity to develop one. A potential solution to this is for facilities to partner with microbiology laboratories and local public health departments to create regional and/or local antibiograms in order to have a more robust idea of local resistance and therefore have a better choice on empiric antibiotic [21–23]. This will potentially decrease the use of broad-spectrum antibiotics and improve patient care through adequate selection up front.

Antibiograms can also be useful to create practice guidelines and/or disease-specific guidelines. Clinicians can create pathways for certain disease states, like urinary tract infection, skin and soft tissue infection, and respiratory bacterial infections by using the local or regional antibiogram coupled with national clinical practice guidelines and pharmacy formulary [24]. Clinical decision support targeting antibiotic prescribing for common conditions along with patient education materials has proven to reduce inappropriate antibiotic prescriptions, decrease the use of broad-spectrum antibiotics, and increase adherence to clinical practice guidelines [25–27].

Measuring outpatient antibiotic prescribing data is used to better understand trends, identify where interventions might be needed to improve prescribing, prioritize efforts, and measure progress or obstacles over time. There is no standard for tracking antibiotic use in the outpatient setting. According to the CDC, at least 30% of antibiotic prescribing in the outpatient is unnecessary with 50% of all outpatient antibiotic use being inappropriate, including unnecessary use, inappropriate selection, dosing and duration [20]. This can also vary by geographic region or state [28]. For this reason, tracking and reporting overall antibiotic prescribing, antibiotic prescribing for specific infections, and agent selection and duration should be part of the AS team efforts in the outpatient setting. Different methods can be used to monitor antibiotic prescribing which include population, visit, prescriber, and indication. Antibiotic prescribed by provider can be linked to patient and indication and subsequently compared to the agents recommended for use in published and institutional guidelines. If the antibiotic prescribed does not correlate with guidelines, this might be considered inappropriate when reported, and feedback could be given to prescriber [29]. Summary reports can be done for entire clinics or health systems down to individual providers and can be stratified by indication, if feasible. This should be used to identify opportunities for improvement, create targeted educational strategies, design interventions and implement policy, track success of interventions, and assess antibiotic use over time. Prescriber audit and feedback and peer comparison in antibiotic prescribing has been proven to be effective in decreasing prescription rates in the outpatient setting. But typically, this is most effective if the intervention is continuous, thus making sustainability difficult [30, 30, 32].

Another important piece of AS that is often taken for granted is adequate documentation. This has been proven lacking in the outpatient setting. Improvement of documentation can be a way to improve antibiotic prescribing in general [33, 34]. Many times the indication and duration of antibiotics can be unclear. Duration of therapy might be biased by dispensing purposes rather than being compliant with guideline and best practices recommendations for a certain disease state. Encouraging prescribers to include an ICD-10 code or diagnosis and duration on the prescription might allow pharmacists to verify appropriateness of agent, dose, and duration, further enhancing the layers of intervention as stewards.

Computerized decision support and electronic health records (EHR), if available, can become a valuable instrument for AS programs in the outpatient setting [35]. Clinical decision support systems (CDSS) can include electronic alerts, antibiotic order sets, and questions that can guide clinicians into ordering the appropriate tests, antibiotics, doses, and durations for certain infections. If implemented correctly, these tools can be valuable to clinicians in a busy outpatient clinic that might not have access to expertise or specialist near their practice. In order to be effective and ensure compliance, clinician's workflow should be taken into consideration [36]. CDSS tools can also be used for surveillance of antibiotic prescribing practices and assess the impact of AS efforts.

An additional area in the outpatient setting that has been identified as a potential for AS interventions are use of point-of-care (POC) diagnostics for infectious diseases, such as influenza, group A strep (GAS) pharyngitis, human immunodeficiency virus, or hepatitis C infection [37]. Accurate and timely diagnosis provides the means to determine appropriate antimicrobial therapy, control further spread of infectious disease, and monitor the emergence of drug resistance through population-based surveillance programs [38, 39]. Since clinical management of suspected infection often starts with empiric antibiotic treatment while awaiting certain diagnostic tests, POC testing availability in clinics could help reduce inappropriate antibiotic use by reducing diagnostic uncertainty, support non-antibiotic prescription decisions, and reassure patients [39]. However, implementation of POC testing in clinics should take into consideration several factors such as existing practice workflow (i.e., patient waits for results in the office or they require a clinician to call later with results), level of competency needed to perform tests in clinics, education to clinicians on the accuracy and subsequent interpretation of each diagnostic test results, whether the tests are reimbursable or not [40].

POC testing in community pharmacies and ambulatory care setting can be used as an opportunity to influence antibiotic prescribing and enhance AS efforts. Based on a study in 2016, about 10,838 community pharmacies possess a Clinical Laboratories Improvement Act (CLIA) certificate of waiver enabling them to perform POC tests [41]. POC tests can be used at community pharmacies to identify infected patients early and linking them to care. But it could also be used to establish collaborative practice agreements between pharmacists and physicians. These collaborations have been studied and recognized as successful [42], in particular for influenza-like illnesses and GAS pharyngitis [32]. Regulations for implementation of pharmacy-based POC testing may vary between states but if permissible it should be explored.

There are a variety of opportunities and interventions that could be established in the outpatient setting to improve antibiotic prescribing and consumption, reduce antimicrobial resistance, and contribute positively to public health. Examples of such interventions in different outpatient settings are described below.

Primary Care Setting

Acute Respiratory Tract Infections

Acute respiratory tract infections (ARI) represent the largest area of outpatient antibiotic prescribing [43]. These conditions include sinusitis, suppurative otitis media, nonsuppurative otitis media, pharyngitis, viral upper respiratory tract infection, bronchitis and bronchiolitis, asthma and allergy, influenza, and viral and nonviral pneumonia. In 2010–2011, ARI accounted for 221 antibiotic prescriptions per 1000 population [43]. Review of antibiotics for these conditions shows that approximately 50% of antibiotic prescriptions were inappropriate [43, 44]. ARIs are commonly of viral etiology, of which antibiotics have no efficacy. In addition to being inappropriately given any antibiotic, many of these prescriptions were for broad-spectrum agents, namely macrolides [44, 45]. In another review azithromycin was prescribed to approximately a quarter of patients over 50 years of age regardless of ARI diagnosis or viral testing result [44]. Visits for acute bronchitis have been associated with up to 73% receiving an antibiotic prescription [46, 47].

Studies have analyzed utilizing CDSS integrated into the EHR to guide appropriate antibiotic prescribing. Data from two health systems within a network whose members use a shared EHR have evaluated the effectiveness of CDSS. In addition to CDSS, participating clinics also received quarterly feedback reports on rates of total antibiotic use, rates of broad-spectrum agent prescribing, and rate of use of CDSS for ARI diagnoses. The resultant decrease of overall inappropriate antibiotic prescribing was modest, however there was a greater decrease seen in broad-spectrum antibiotic use [45]. In another example of CDSS, the specific use of azithromycin and gatifloxacin for ARI was targeted, resulting in a decrease of their use from 22% to 3.3% for ARI where antibiotics are not warranted [48]. In a follow-up study when gatifloxacin was replaced with moxifloxacin but not included in the CDSS target, the investigators found an increase in non-guideline concordant use of moxifloxacin [49].

Despite clinical practice guidelines [50] and CDC treatment recommendations [51] for the appropriate treatment of ARI, unnecessary prescribing of antibiotics remains prevalent. Studies analyzing behavioral interventions have shown to be effective in reducing inappropriate antibiotic use. Meeker et al. [52] studied the impact of three interventions on antibiotic prescribing for ARI: non-pharmacologic alternative suggestions, written justification for treatment decision in the EHR, or peer-comparison emails where clinicians were either a "top performer" or "not a top performer." Of these methods, peer-comparison and accountable justification led to lower rates of antibiotic prescribing. In a 12-month follow-up assessment after stopping the interventions, inappropriate prescribing for accountable justification and suggested alternatives increased to above control levels, whereas peer-comparison continued to maintain a significant decrease [53]. It should be noted that for all interventions a decrease in inappropriate prescribing was observed, including the control groups.

Urinary Tract Infections

Urinary tract infections (UTIs) are one of the most commonly diagnosed infectious diseases and account for a considerable proportion of outpatient antibiotic use. From 2007 to 2009 in the USA, there were eight million ambulatory care visits for UTIs, 60% of which resulted in an antibiotic prescription [20]. Of those 69% were for a broad-spectrum antibiotic [20]. An analysis of US data from 2002 to 2011 found that up to 80% of UTI-related visits resulted in an antibiotic prescription [54]. The Infectious Diseases Society of America (IDSA) has developed clinical practice guidelines for the management of uncomplicated cystitis and pyelonephritis [55], catheter-associated UTI [56], and asymptomatic bacteriuria (ASB); [57] however, antibiotic prescribing in ASB and UTIs continues to be sub-optimal [54, 58, 59].

First-line recommended antibiotics are trimethoprim-sulfamethoxazole (TMP-SMX) for 3 days, nitrofurantoin for 5 days, or single-dose fosfomycin [55]. Fluoroquinolones are recommended as second-line alternatives due to risk of collateral damage [55]. A review of national ambulatory prescribing patterns for UTI from 1989 to 1999 found that despite no change from TMP-SMX being the recommended first-line antibiotic, there was a shift toward increased fluoroquinolone (FQ) and nitrofurantoin use, 19% to 29% and 14% to 30%, respectively, with a decline in TMP-SMX prescribing from 48% to 24% [59]. In a similar review that analyzed UTI prescribing from 2002 to 2011, the authors determined that FQs accounted for 49% of antibiotic prescriptions, TMP-SMX 27%, and nitrofurantoin 19% [54]. In an analysis of predictors of antimicrobial choice, both reviews found that older age and primary care by an internist to be associated with FQ use [54, 59]. Durkin et al. looked at 2011 IDSA guideline adherence for uncomplicated UTI and found that FQ use accounted for 43% of prescriptions and was the most commonly prescribed antibiotic both before and after guideline publication [60]. In a study that looked at prescribing in an internal medicine clinic, only 55% of prescriptions for UTIs were for a first-line agent [61].

FQ use has generated significant conversation in regard to AS. In addition to being second-line alternative agents for UTI [55], the risk of serious adverse effects of FQs has been the focus of several warnings from the Food and Drug Administration (FDA) [61]. In May 2016, the FDA advised that the risk of FQ use may outweigh the benefit "for patients with … uncomplicated urinary tract infections who have other treatment options." [62] Compared to low-risk antibiotics, ciprofloxacin use in UTI has been associated with a nearly threefold risk of *Clostridioides difficile* infection [63], which is only compounded in patients who have a history of gastrointestinal disease. The broad-spectrum activity, high oral bioavailability, and convenient dosing regimens have led to wide-spread use not limited to UTIs. Consequences of FQ use include infection with methicillin-resistant *Staphylococcus aureus* and development of resistance in other Gram negative bacilli, such as *Pseudomonas aeruginosa* [55].

A significant limitation of much of the available data on appropriate prescribing in UTI is the lack of allergy and/or susceptibility information, which may necessitate a second-line agent. The IDSA guidelines recommend that TMP-SMX should be avoided if local resistance is >20%, which may be the case in regions [55]. Nitrofurantoin is not recommended for patients with a creatinine clearance <30 mL/min [64] and has been listed as a high-risk medication for older adults [65], though has been revised to specify when used for >90 days [64, 65].

In addition to antibiotic selection, prescribed durations are also often discordant with guideline recommendations. In a review of guideline adherence for UTIs, most antibiotics were prescribed for 3-, 5-, 7-, or 10-day durations, the majority of which were for 7 days [60]. The appropriate duration occurred in 21% of FQ, 25% of TMP-SMX, and 16% of nitrofurantoin prescriptions, and overall >75% of durations were inappropriate. In another analysis, ciprofloxacin was most commonly prescribed for a 10-day course, despite the guideline recommendation of 3 days [63].

ASB in particular remains a significant challenge to AS. By definition ASB is the absence of UTI signs or symptoms with a urine culture growing at least 10^5 colony-forming units of 1 or more species of bacteria [57]. With notable exceptions for pregnant patients and those undergoing urologic procedures involving mucosal trauma, the majority of patients should be neither screened nor treated for ASB because of absence of evidence showing benefit with antibiotic treatment [57]. Obtaining urine cultures when not clinically indicated can lead to inappropriate treatment of ASB due to a positive result [66]. Despite the recommendation to avoid antibiotics, 45% [67] to 72% [68] of ASB cases are treated.

Impediments to AS remain around the selection and duration of antibiotics for UTI and avoidance of treatment of ASB despite updates to clinical practice guidelines. Antibiotic prescribing was largely unchanged following the 2010 UTI guideline update that stressed FQs as second-line agents [55, 60]. An alternative approach included in the United Kingdom's NICE guideline for lower UTI is to delay treatment in some non-pregnant women with acute cystitis, as symptoms may resolve on their own (NICE 2020) [69]. Additional stewardship initiatives need to be implemented at a facility or clinic level. Two such programs [70, 71] where guideline-based algorithms were implemented in emergency departments increased appropriate antibiotic prescribing. An electronic UTI order set improved guideline concordance from 44% to 84% [70]. The creation of a treatment selection tool and best practice algorithm resulted in a significant decrease of ciprofloxacin prescribing from 32% to 11% [71]. Similar interventions may benefit other ambulatory care settings.

Outpatient Parenteral Antimicrobial Therapy

Outpatient parenteral antimicrobial therapy (OPAT) allows for administration of antimicrobials that traditionally had been administered within an inpatient setting and offers many benefits to patients, including shortened hospital stays and the potential to avoid acquisition of additional infections [72]. Studies have cited that 2–2.5% of adult and pediatric hospital discharges [73, 74] utilize OPAT for a wide range of indications, such as bone and joint infections, urinary tract infections, skin/soft tissue infections, etc. [74] Evaluation of the antimicrobial regimens utilized in OPAT cases reveals that there is potentially room for improvement in antibiotic prescribing in this setting. One study in a pediatric setting found that 56% of regimens had a definite opportunity for intervention [75]; a similar percentage of adult OPAT regimens had opportunities for interventions [74].

One concerning observation seen in studies evaluating OPAT regimens is that there are many instances in which OPAT was utilized when potentially not needed. Many studies identify OPAT regimens for highly bioavailable antimicrobials that could have been prescribed orally. One study [76] identified that 34% of evaluated regimens could have potentially been switched to an oral regimen; many other studies have also identified this as an area for improvement when selecting OPAT regimens [74, 75, 77–79]. Additionally, several studies [74, 75, 77–80] identified a smaller proportion of prescribed regimens (up to 14%) [78] without an indication for continuation of therapy. Other common opportunities for intervention involved antibiotic selection (both in spectrum of activity and antibiotics not matching causative pathogen), alterations in duration of therapy, as well as laboratory monitoring [74, 75, 77, 81–83]. Several studies also identified a role for pharmacists to intervene to improve quality of care, as many of the identified areas for improvement involved antimicrobial dosing and follow-up laboratory monitoring (such as aminoglycoside levels) [82].

Thus, the literature has shown that there is a clear role for AS in the setting of OPAT. While OPAT is administered by definition in an outpatient setting, studies published examining effect of AS have focused on intervening prior to the point of patient transition to the outpatient setting. The commonality among these studies is that there is a review for appropriateness prior to patient discharge into the outpatient setting—the main differences among the studies lie in who is performing the audit. There are studies that focus on the effects of the impact of a mandatory infectious disease consultation to approve OPAT [79, 84]. These studies were able to successfully demonstrate that an infectious disease specialist was able help improve OPAT utilization; one study [84] showed that 38.6% of consultations resulted in switch from an intravenous (IV) to oral antimicrobial, while 29.6% of patients had their parenteral antibiotic changed to another agent. A small number of patients also had antibiotic doses and durations of therapy amended, and there were a small percentage of patients identified that did not need any further antibiotic therapy. Another study [79] comparing outcomes of patients in whom OPAT was denied by an infectious disease consultant against those who received OPAT found that the infectious diseases consultant was able to recommend interventions similar to those seen elsewhere in the literature. The literature also shows that such interventions do not worsen patient outcomes and are therefore safe, and at the same time, also result in a substantial cost savings [79, 84–86].

Other studies chose to evaluate the impact of a multidisciplinary stewardship team on OPAT and were also able to demonstrate a meaningful impact on OPAT use. One such study [78] examined effects of a stewardship team consisting of a physician and a pharmacist in a pediatric hospital. This team was able to decrease overall OPAT usage from 7.7% of discharged patients to 6.1% of discharged patients ($p < 0.01$), as well as make recommendations on a large proportion of patients being discharged with OPAT (such as recommending alternative antibiotics and switching to an orally administered antibiotic). This study [78] was also able to demonstrate that implementation of the team's recommendations was safe, as evidenced by lower rates of OPAT complications and no change in readmissions in patients switched to an oral antimicrobial. Another study [77] utilized a larger team, involving a pharmacist, a physician, and a case manager. This team was also able to make a meaningful impact in OPAT utilization; OPAT was avoided in 13.2% of cases and a significant cost savings was seen.

Safety of OPAT interventions has also been documented within the literature. Several studies examined failure rates of de-escalated antibiotic regimens and found that de-escalated regimens had similar clinical success rates as compared to regimens without stewardship interventions [79, 84, 85]. One study even found that review of an OPAT regimen by an infectious diseases specialist resulted in lower readmission rates within 30 days, as well as significantly fewer emergency department visits post discharge [86], thus suggesting that OPAT stewardship interventions could lead to fewer clinical failures. One study found that increased rates of toxicity from OPAT regimens within the group reviewed by an infectious diseases expert, however this group also had significantly more patients with laboratory monitoring being performed; thus, the authors concluded that toxicity was severely underestimated within the group that was not reviewed by an infectious diseases specialist [81]. Other studies confirmed that ordering of laboratory monitoring was a common safety intervention where OPAT stewardship programs are able to intervene [73, 77]. Less information is available about impact of OPAT stewardship interventions on catheter-associated complications; one study [78] did not show any difference in catheter-related complications with stewardship interventions, but other literature suggests that OPAT stewardship interventions may help prevent catheter placement [77] and could possibly help limit complications associated with catheters [73].

Outpatient Hemodialysis Centers

Patients receiving chronic hemodialysis are uniquely predisposed to acquisition of infections for multiple reasons. The process of hemodialysis itself provides opportunities for introduction of infection-causing pathogens via possible contamination of water supply, distribution systems, hemodialysis machines, etc. [87] Frequent vascular access associated with hemodialysis also provides frequent opportunities for introduction of pathogens directly into the bloodstream [87]. Additionally, patients with end-stage renal disease have compromised immune systems as a result of not only immune changes associated with uremia, but also potentially due to underlying etiology of renal failure, such as diabetes or auto-immune disease [88]. In addition to being at increased risk of developing infections, they are also at increased risk of acquisition of a multi-drug resistant organism [89], which can lead to increases in both morbidity [90] and mortality [91]. Thus, there is a need to help preserve available antibiotic options to help minimize these risks via AS. One of the first attempts to quantify antimicrobial use in outpatient dialysis facilities was based on data submitted to the CDC National Healthcare Safety Network (NHSN) [92]. Information on hospital admissions, outpatient initiation of an IV antimicrobial, or a positive blood culture was provided to the NHSN from 32 outpatient dialysis centers. A total of 977 instances of antimicrobial starts were reported for the year, with reported rates of use varying based on vascular access availability, ranging from 1.8 per 100 patient months in those with fistulas up to 25.4 per 100 patient months in those with temporary central lines. Vancomycin accounted for almost three-quarters of the antimicrobials initiated [92]. Another study by George et al. [93] provided an estimated rate of IV antibiotic initiation of 7.7 per 100 patient months. These studies, however, only quantified antibiotic initiation events and did not provide any information about appropriateness of orders or about antibiotic consumption or duration.

A study done by Snyder et al. [94] aimed to address and quantify appropriateness of antibiotic usage, as well as to better characterize antibiotic consumption in two outpatient dialysis centers in Boston, Massachusetts. This study found that over a 1-year period, 32% of studied patients received at least one IV antibiotic dose, resulting in a rate of antibiotic use of 39.3 doses per patient month [94]. When reviewed for appropriateness, 29.8% of doses were deemed to be inappropriate, half of which were deemed so due to lack of infection. Further analysis of the data showed that risk factors for antimicrobial receipt included tunneled catheter access, history of colonization or infection with a multi-drug resistant organism, and receipt of dialysis treatments during a day-time shift [94]. The latter risk factor was surmised to be due to the fact that patients receiving treatment during the evening were less likely to be empirically treated on the unit and were more likely to be evaluated in an emergency setting. Risk factors for receipt of inappropriate antimicrobials differed slightly, but still included presence of a tunneled catheter, as well as longer duration of chronic hemodialysis [95]. Receipt of dialysis treatments in one particular dialysis unit was also identified as a risk factor for receipt of inappropriate antibiotic use, thus implying that differences in practice between individual providers play a role in prescribing practices [96]. This underscores the importance of individual dialysis centers having an understanding of their antibiotic usage patterns to best target areas for intervention.

Another study done in Australia [97] also aimed to quantify antibiotic consumption and appropriateness of antibiotic use in outpatient dialysis centers. In addition to examining IV antibiotic usage, Hui et al. [97] also examine oral antibiotic usage.

This study found that 55.3% of enrolled patients received an antimicrobial during the 6-month study period. This study [97] found 24.6% of antibiotic regimens to be inappropriate. Incorrect antibiotic dose or frequency was the most common reason regimens were deemed inappropriate (78.2% of regimens); other reasons for regimens deemed to be inappropriate were incorrect duration (9.1%), allergy mismatches (7.3%), inappropriate spectrum of activity (7.3% too narrow, 5.5% too broad), unnecessary therapy (5.5%), microbiology mismatch (3.6%), and inappropriate duration of surgical prophylaxis (1.8%) [97].

The aforementioned studies [95, 97] show that there is ample room for improvement in antibiotic prescribing in outpatient dialysis facilities. Despite the above data, there is very little information available in the literature describing implementation of a stewardship program in this setting. A modeling study done by D'Agata et al. [98] projected that a stewardship program with the aim of reducing inappropriate antibiotic use by 20% could make a significant impact. A 4.8% reduction in infections caused by multi-drug resistant organisms and *Clostridium difficile* is predicted, as well as a 4.6% reduction in infection-related deaths [98]. There is also a significant cost savings associated with such a program; the model estimates a 5% cost savings per year, which translates into a savings of over $100,000,000 per year [98].

There is one study available that describes the impact of implementation of an AS program in a group of 6 outpatient dialysis facilities [99]. The program utilized a multifaceted approach that relied on leadership support, education for all staff members (including physicians, nurses, nurse practitioners, physician assistants, dialysis technicians, dieticians, and social workers), and regular conference calls with leadership from all six sites to review antibiotic usage and discuss how to help optimize usage at each site. Additionally, the program had a focus on involving staff at each facility to help implement the stewardship program and the associated cultural changes that would be necessary for the program to be successful. At the end of the study period, mean monthly rate of antimicrobial doses fell from 22.6 doses per 100 patient months to 10.5 doses per 100 patient months [99]. During the conference calls, changes to antimicrobial therapy were recommended in 20.6% of courses that were discussed in sufficient detail. The most common recommendations made were to de-escalate antimicrobial therapy to more narrow-spectrum agents and to discontinue therapy when criteria for infection were not met. Implementation of this program was deemed to be safe, as there were no changes seen in rate of hospital admission, blood stream infections, or composite of facility-level quality measures after implementation of the stewardship program [99].

Conclusion

AS in the outpatient setting is essential in the efforts to tackle the misuse of antibiotics, halt the development of antimicrobial resistance, and improve healthcare quality and patient safety throughout the healthcare continuum. Risks associated with antibiotic consumption pertain to not only antibiotic resistance, but also antibiotic-related adverse events, such as drug reactions, life-threatening allergic reactions, and *Clostridioides difficile* infections [100, 101]. Antimicrobial resistance is no longer limited to the hospital setting but is unfortunately now found in the community setting.

Implementing a comprehensive AS program in the outpatient setting is challenging. The evidence on AS programs effectiveness on measurable patient or resistance outcomes is lacking, making stewardship efforts less compelling. There are several factors that would directly impact the successful implementation of a successful AS program, such as limited resources; unawareness of costs; healthcare provider, patient, and family lack of knowledge; pressure of patient satisfaction; and struggles with scalability and feasibility. AS initiatives must be tailored to the community, individual practices, and healthcare systems in order to be effective. Healthcare providers must be committed to use antibiotics judiciously in clinical practice. Partnerships with experts in the field and public health officials can help break down knowledge gaps, generate opportunities for available resources, and promote coordinated approach to AS efforts [102].

It has become evident that antibiotic resistance is a worldwide public health crisis and that the only way to stop its dissemination is by joint efforts not only locally or nationally but globally. Coordination of multiple healthcare systems in a multidisciplinary approach with collaboration of public health agencies might help narrow the gap in resources and spread the word more efficiently.

References

1. Centers for Disease Control and Prevention. Making health care safer: stop spread of antibiotic resistance; 2017. https://www.cdc.gov/vitalsigns/stop-spread/index.html. Accessed 17 May 2019.
2. Suda KJ, Hicks LA, Roberts RM, Hunkler RJ, Danziger LH. A national evaluation of antibiotic expenditures by healthcare setting in the United States, 2009. J Antimicrob Chemother. 2013;68(3):715–8.
3. Hurley HJ, Knepper BC, Price CS, Mehler PS, Burman WJ, Jenkins TC. Avoidable antibiotic exposure for uncomplicated skin and soft tissue infections in the ambulatory care setting. Am J Med. 2013;126(12):1099–106.
4. National Action Plan for Combating Antibiotic-Resistant Bacteria; 2015. https://obamawhitehouse.archives.gov/sites/default/files/docs/national_action_plan_for_combating_antibotic-resistant_bacteria.pdf. Accessed 8 June 2019.
5. Sanchez GV, Fleming-Dutra KE, Roberts RM, Hicks LA. Core elements of outpatient antibiotic stewardship. MMWR Recomm Rep. 2016;65(RR-6):1–12.
6. Drekonja D, Filice G, Greer N, Olson A, MacDonald R, Rutks I, Wilt T. Antimicrobial stewardship in outpatient settings: a systematic review. Infect Control Hosp Epidemiol. 2015;36(2):142–52. https://doi.org/10.1017/ice.2014.41.
7. Dellit TH, Owens RC, McGowan JE Jr, Gerding DN, Weinstein RA, Burke JP, et al. Infectious Diseases Society of America and the Society for Healthcare Epidemiology of America guidelines for developing an institutional program to enhance antimicrobial stewardship. Clin Infect Dis. 2007;44(2):159–77. https://doi.org/10.1086/510393.
8. Pollack LA, van Santen KL, Weiner LM, Dudeck MA, Edwards JR, Srinivasan A. Antibiotic stewardship programs in U.S. acute care hospitals: findings from the 2014 National Healthcare Safety Network Annual Hospital Survey. Clin Infect Dis. 2016;63(4):443–9.

9. Olans RN, Olans RD, DeMaria A. The critical role of the staff nurse in antimicrobial stewardship—unrecognized, but already there. Clin Infect Dis. 2016;62(1):84–9. https://doi.org/10.1093/cid/civ697.
10. Centers for Disease Control and Prevention and The American Nurses Association. Redefining the Antibiotic Stewardship Team: recommendations from the American Nurses Association/Centers for Disease Control and Prevention Workgroup on the role of registered nurses in hospital antibiotic stewardship practices. Available from: https://www.cdc.gov/antibiotic-use/healthcare/pdfs/ANA-CDC-whitepaper.pdf.
11. Edwards R, Drumright L, Kiernan M, Holmes A. Covering more territory to fight resistance: considering nurses' role in antimicrobial stewardship. J Infect Prev. 2011;12(1):6–10.
12. Meeker D, Knight TK, Friedberg MW, et al. Nudging guideline-concordant antibiotic prescribing: a randomized clinical trial. JAMA Intern Med. 2014;174(3):425–31.
13. Belongia EA, Knobloch MJ, Kieke BA, Davis JP, Janette C, Besser RE. Impact of statewide program to promote appropriate antimicrobial drug use. Emerg Infect Dis. 2005;11(6):912–20. https://doi.org/10.3201/eid1106.050118.
14. Butler CC, Simpson SA, Dunstan F, Rollnick S, Cohen D, Gillespie D, et al. Effectiveness of multifaceted educational programme to reduce antibiotic dispensing in primary care: practice based randomised controlled trial. BMJ. 2012;344:d8173.
15. Finkelstein JA, Huang SS, Kleinman K, Rifas-Shiman SL, Stille CJ, Daniel J, et al. Impact of a 16-community trial to promote judicious antibiotic use in Massachusetts. Pediatrics. 2008;121(1):e15–23.
16. Gonzales R, Steiner JF, Lum A, Barrett PH Jr. Decreasing antibiotic use in ambulatory practice: impact of a multidimensional intervention on the treatment of uncomplicated acute bronchitis in adults. JAMA. 1999;281(16):1512–9.
17. Little P, Stuart B, Hobbs FD, et al. Antibiotic prescription strategies for acute sore throat: a prospective observational cohort study. Lancet Infect Dis. 2014;14(3):213–9.
18. Little P, Rumsby K, Kelly J, et al. Information leaflet and antibiotic prescribing strategies for acute lower respiratory tract infection: a randomized controlled trial. JAMA. 2005;293(24):3029–35.
19. Chao JH, Kunkov S, Reyes LB, Lichten S, Crain EF. Comparison of two approaches to observation therapy for acute otitis media in the emergency department. Pediatrics. 2008;121(5):e1352–6.
20. Shapiro DJ, Hicks LA, Pavia AT, Hersh AL. Antibiotic prescribing for adults in ambulatory care in the USA, 2007-09. J Antimicrob Chemother. 2014;69(1):234–40.
21. Kohlmann R, Gatermann SG. Analysis and presentation of cumulative antimicrobial susceptibility test data – the influence of different parameters in a routine clinical microbiology laboratory. PLoS One. 2016;11(1):e0147965. https://doi.org/10.1371/journal.pone.0147965.
22. Saxena S, Ansari SK, Raza MW, Dutta R. Antibiograms in resource limited settings: are stratified antibiograms better? Infect Dis. 2016;48(4):299–302. https://doi.org/10.3109/23744235.2015.1113437.
23. McGregor JC, Bearden DT, Townes JM, et al. Comparison of antibiograms developed for inpatients and primary care outpatients. Diagn Microbiol Infect Dis. 2013;76(1):73–9. https://doi.org/10.1016/j.diagmicrobio.2013.01.026.
24. Barlam TF, Cosgrove SE, Abbo LM, MacDougall C, Schuetz AN, Septimus EJ, et al. Implementing an Antibiotic Stewardship program: guidelines by the Infectious Diseases Society of America and the Society for Healthcare Epidemiology of America. Clin Infect Dis. 2016;62(10):e51–77. https://doi.org/10.1093/cid/ciw118.
25. McGinn TG, et al. Efficacy of an evidence-based clinical decision support in primary care practices: a randomized clinical trial. JAMA Intern Med. 2013;173(17):1584–11.
26. Jenkins TC, et al. Effects of clinical pathways for common outpatient infections on antibiotic prescribing. Am J Med. 2013;126(4):327–35 e312.
27. Forrest CB, et al. Improving adherence to otitis media guidelines with clinical decision support and physician feedback. Pediatrics. 2013;131(4):e1071–81.

28. Hicks LA, Bartoces MG, Roberts RM, et al. US outpatient antibiotic prescribing variation according to geography, patient population, and provider specialty in 2011. Clin Infect Dis. 2015;60(9):1308–16.
29. Klepser ME, Altom A, MacCallum C, Nordquist E, Klepser DG. Description of the methods for describing and assessing the appropriateness of antibiotic prescribing and adherence to published treatment guidelines in an academic medical clinic. Innov Pharm. 2016;7(1):Article 2.
30. Gerber JS, Prasad PA, Fiks AG, et al. Effect of an outpatient antimicrobial stewardship intervention on broad-spectrum antibiotic prescribing by primary care pediatricians: a randomized trial. JAMA. 2013;309(22):2345–52.
31. Gerber JS, Prasad PA, Fiks AG, et al. Durability of benefits of an outpatient antimicrobial stewardship intervention after discontinuation of audit and feedback. JAMA. 2014;312(23):2569–70.
32. Thornley T, Marshall G, Howard P, Wilson AP. A feasibility service evaluation of screening and treatment of group A streptococcal pharyngitis in community pharmacies. J Antimicrob Chemother. 2016;71(11):3293–9. https://doi.org/10.1093/jac/dkw264.
33. Klepser ME, Dobson EL, Pogue JM, Labreche MJ, Adams AJ, Gauthier TP, et al. A call to action for outpatient antibiotic stewardship. J Am Pharm Assoc. 2017;57(4):457–63.
34. Jones BE, Sauer B, Jones MM, et al. Variation in outpatient antibiotic prescribing for acute respiratory infections in the veteran population: a cross-sectional study. Ann Intern Med. 2015;163(2):73–80.
35. Kullar R, Goff DA, Schulz LT, Fox BC, Rose WE. The “epic” challenge of optimizing antimicrobial stewardship: the role of electronic medical records and technology. Clin Infect Dis. 2013;57(7):1005–13. https://doi.org/10.1093/cid/cit318.
36. Forrest GN, Van Schooneveld TC, Kullar R, Schulz LT, Duong P, Postelnick M. Use of electronic health records and clinical decision support systems for antimicrobial stewardship. Clin Infect Dis. 2014;59(Suppl 3):S122–33. https://doi.org/10.1093/cid/ciu565.
37. Dobson EL, Klepser ME, Pogue JM, Labreche MJ, Adams AJ, Gauthier TP, et al. Outpatient antibiotic stewardship: interventions and opportunities. J Am Pharm Assoc. 2017;57(4):464–73.
38. AdvaMedDx. Global commitment on diagnostic tests to fight antimicrobial resistance. Statement, Mar 17, 2017.
39. Shulman ST, Bisno AL, Clegg HW, Gerber MA, Kaplan EL, Lee G, et al. Clinical practice guideline for the diagnosis and management of group a streptococcal pharyngitis: 2012 update by the Infectious Diseases Society of America. Clin Infect Dis. 2012;55(10):e86–e102. https://doi.org/10.1093/cid/cis629.
40. Avent ML, Cosgrove SE, Price-Haywood EG, et al. Antimicrobial stewardship in the primary care setting: from dream to reality? BMC Fam Pract. 2020;21:134. https://doi.org/10.1186/s12875-020-01191-0.
41. Klepser ME, Adams AJ, Srnis P, Mazzucco M, Klepser D. U.S. community pharmacies as CLIA-waived facilities: prevalence, dispersion, and impact on patient access to testing. Res Social Adm Pharm. 2016;12(4):614–21.
42. Klepser ME, Klepser DG, Dering-Anderson AM, Morse JA, Smith JK, Klepser SA. Effectiveness of a pharmacist-physician collaborative programto manage influenza-like illness. J Am Pharm Assoc. 2016;56(1):14–21.
43. Fleming-Dutra KE, Hersh AL, Shapiro DJ, Bartoces M, Enns EA, et al. Prevalence of inappropriate antibiotic prescriptions among US ambulatory care visits, 2010-2011. JAMA. 2016;315(17):1864–73.
44. Havers FP, Hicks LA, Chung JR, Gaglani M, Murthy K, et al. Outpatient antibiotic prescribing for acute respiratory infections during influenza seasons. JAMA Netw Open. 2018;1(2):e180243.
45. Litvin CB, Ornstein SM, Wessell AM, Nemeth LS, Niertert PJ. Use of an electronic health record clinical decision support tool to improve antibiotic prescribing for acute respiratory infections: the ABX-TRIP study. J Gen Intern Med. 2013;28(6):810–6.

46. Barlam TF, Morgan JR, Wetzler LM, Christiansen CL, Drainoni ML. Antibiotics for respiratory tract infections: a comparison of prescribing in an outpatient setting. Infect Control Hosp Epidemiol. 2015;36(2):153–9.
47. Barnett ML, Linder JA. Antibiotic prescribing in adults with sore throat in the United States, 1997-2019. JAMA Intern Med. 2014;174:138–40.
48. Rattinger GB, Mullins CD, Zuckerman IH, Onukwugha E, Walker LD, et al. A sustainable strategy to prevent misuse of antibiotics for acute respiratory infections. PLoS One. 2012;7(12):e51147.
49. Gifford J, Vaeth E, Richards K, Siddiqui T, Gill C, Wilson L, DeLisle S. Decision support during electronic prescription to stem antibiotic overuse for acute respiratory infections: a long-term, quasi-experimental study. BMC Infect Dis. 2017;17(1):528.
50. Wong DM, Blumberg DA, Lowe LG. Guidelines for the use of antibiotics in acute upper respiratory tract infections. Am Fam Physician. 2006;74(956–66):969.
51. Centers for Disease Control and Prevention. Adult treatment recommendations; 2017. https://www.cdc.gov/antibiotic-use/community/for-hcp/outpatient-hcp/adult-treatment-rec.html. Accessed 26 May 2019.
52. Meeker D, Linder JA, Fox CR, Friedberg MW, Persell SD, et al. Effect of behavioral interventions on inappropriate antibiotic prescribing among primary care practices: a randomized clinical trial. JAMA. 2016;315(6):562–70.
53. Linder JA, Meeker D, Fox CR. Effects of behavioral interventions on inappropriate antibiotic prescribing in primary care 12 months after stopping interventions. JAMA. 2017;318(14):1391–2.
54. Kobayashi M, Shapiro DJ, Hersh A, Sanchez GV, Hicks LA. Outpatient antibiotic prescribing practices for uncomplicated urinary tract infection in women in the United States, 2002-2011. Open Forum Infect Dis. 2016;3(3):ofw159.
55. Gupta K, Hooton TM, Naber KG, Wullt B, Colgan R, et al. International clinical practice guidelines for the treatment of acute uncomplicated cystitis and pyelonephritis in women: a 2010 update by the Infectious Diseases Society of America and the European Society for Microbiology and Infectious Diseases. Clin Infect Dis. 2011;52(5):e103–20.
56. Hooton TM, Bradley SF, Cardenas DD, Colgan R, Geerlings SE, et al. Diagnosis, prevention, and treatment of catheter-associated urinary tract infection in adults: 2009 international clinical practice guidelines from the Infectious Diseases Society of America. Clin Infect Dis. 2010;50(5):625–63.
57. Nicolle LE, Gupta K, Bradley SF, Colgan R, DeMuri GP, et al. Clinical practice guideline for the management of asymptomatic bacteriuria: 2019 update by the Infectious Diseases Society of America. Clin Infect Dis. 2019;68(10):e83–e110.
58. Shively NR, Buehrle DJ, Clancy CJ, Decker BK. Prevalence of inappropriate antibiotic prescribing in primary care clinics within a veterans affairs health care system. Antimicrob Agents Chemother. 2018;62(8):e00337–e003318.
59. Huang ES, Stafford RS. National patterns in the treatment of urinary tract infections in women by ambulatory care physicians. Arch Intern Med. 2002;162:41–7.
60. Durkin MJ, Keller M, Butler AM, Kwon JH, Dubberke ER, et al. An assessment of inappropriate antibiotic use and guideline adherence for uncomplicated urinary tract infection. Open Forum Infect Dis. 2018;5(9):ofy198.
61. US Food and Drug Administration. FDA drug safety communication: FDA updates warnings for oral and injectable fluoroquinolone antibiotics due to disabling side effects; 2018. https://www.fda.gov/drugs/drug-safety-and-availability/fda-drug-safety-communication-fda-updates-warnings-oral-and-injectable-fluoroquinolone-antibiotics. Accessed 25 May 2019.
62. US Food and Drug Administration. FDA drug safety communication: FDA advises restricting fluoroquinolone antibiotic use for certain uncomplicated infections; warns about disabling side effects that can occur together; 2018. https://www.fda.gov/drugs/drug-safety-and-availability/fda-drug-safety-communication-fda-advises-restricting-fluoroquinolone-antibiotic-use-certain. Accessed 25 May 2019.

63. Ge IY, Fevrier HB, Conell C, Kheraj MN, Flint AC, Smith DS, Herrinton LJ. Reducing risk of Clostridium difficile infection and overall use of antibiotic in the outpatient treatment of urinary tract infection. Ther Adv Urol. 2018;10(10):283–93.
64. The American Geriatrics Society 2015 Beers Criteria Update Expert Panel. American Geriatrics Society 2015: updated beers criteria for potentially inappropriate medication use in older adults. J Am Geriatr Soc. 2015;63:2227–46.
65. Sprague L. The star rating system and medicare advantage plans. Issue Brief Natl Health Policy Forum 2015. https://www.nhpf.org/library/details.cfm/2988. Accessed 25 May 2019.
66. Sloane PD, Kistler CE, Reed D, Weber DJ, Ward K, Zimmerman S. Urine culture testing in community nursing homes: gateway to antibiotic overprescribing. Infect Control Hosp Epidemiol. 2017;38(5):524–31. https://doi.org/10.1017/ice.2016.326.
67. Flokas ME, Andreatos N, Alevizakos M, Kalbasi A, Onur P, Mylonakis E. Inappropriate management of asymptomatic patients with positive urine cultures: a systematic review and meta-analysis. Open Forum Infect Dis. 2017;4(4):ofx207. https://doi.org/10.1093/ofid/ofx207.
68. Spivak ES, Burk M, Zhang R, et al. Management of bacteriuria in veterans affairs hospitals. Clin Infect Dis. 2017;65(6):910–7. https://doi.org/10.1093/cid/cix474.
69. National Institute for Health and Care Excellence. Urinary tract infection (lower): antimicrobial prescribing. [NICE Guideline No. 109]; 2018. https://www.nice.org.uk/guidance/ng109.
70. Hecker MT, Fox CJ, Son AH, et al. Effect of stewardship intervention on adherence to uncomplicated cystitis and pyelonephritis guidelines in an emergency department setting. PLoS One. 2014;9:387899.
71. Landry E, Sulz L, Bell A, Rathgeber L, Balogh H. Urinary tract infections: leading initiatives in selecting empiric outpatient treatment (UTILISE). Can J Hosp Pharm. 2014;67(2):116–25.
72. Norris AH, Shrestha NK, Allison GM, et al. 2018 Infectious Diseases Society of America clinical practice guideline for the management of outpatient parenteral antimicrobial therapy. Clin Infect Dis. 2019;68(1):e1–e35.
73. Madigan T, Banerjee R. Characteristics and outcomes of outpatient parenteral antimicrobial therapy at an academic children's hospital. Pediatr Infect Dis J. 2013;32(4):346–9.
74. Spivak ES, Kendall B, Orlando P, et al. Evaluation of outpatient parenteral antimicrobial therapy at a veterans affairs hospital. Infect Control Hosp Epidemiol. 2015;36(9):1103–5.
75. Knackstedt ED, Stockmann C, Davis CR, Thorell EA, Pavia AT, Hersh AL. Outpatient parenteral antimicrobial therapy in pediatrics: an opportunity to expand antimicrobial stewardship. Infect Control Hosp Epidemiol. 2015;36(2):222–4.
76. Goldman JL, Richardson T, Newland JG, et al. Outpatient parenteral antimicrobial therapy in pediatric medicaid enrollees. J Pediatr Infect Dis Soc. 2017;6(1):65–71.
77. Heintz BH, Halilovic J, Christensen CL. Impact of a multidisciplinary team review of potential outpatient parenteral antimicrobial therapy prior to discharge from an academic medical center. Ann Pharmacother. 2011;45(11):1329–37.
78. Hersh AL, Olson J, Stockmann C, et al. Impact of antimicrobial stewardship for pediatric outpatient parenteral antibiotic therapy. J Pediatr Infect Dis Soc. 2018;7(2):e34–6.
79. Conant MM, Erdman SM, Osterholzer D. Mandatory infectious diseases approval of outpatient parenteral antimicrobial therapy (OPAT): clinical and economic outcomes of averted cases. J Antimicrob Chemother. 2014;69(6):1695–700.
80. Dryden M, Saeed K, Townsend R, et al. Antibiotic stewardship and early discharge from hospital: impact of a structured approach to antimicrobial management. J Antimicrob Chemother. 2012;67(9):2289–96.
81. Akar A, Singh N, Hyun DY. Appropriateness and safety of outpatient parenteral antimicrobial therapy in children: opportunities for pediatric antimicrobial stewardship. Clin Pediatr. 2014;53(10):1000–3.
82. Hodgson KA, Huynh J, Ibrahim LF, et al. The use, appropriateness and outcomes of outpatient parenteral antimicrobial therapy. Arch Dis Child. 2016;101(10):886–93.
83. Wijnakker R, Visser LE, Schippers EF, Visser LG, van Burgel ND, van Nieuwkoop C. The impact of an infectious disease expert team on outpatient parenteral antimicrobial treatment in the Netherlands. Int J Clin Pharm. 2019;41(1):49–55.

84. Sharma R, Loomis W, Brown RB. Impact of mandatory inpatient infectious disease consultation on outpatient parenteral antibiotic therapy. Am J Med Sci. 2005;330(2):60–4.
85. Shrestha NK, Bhaskaran A, Scalera NM, Schmitt SK, Rehm SJ, Gordon SM. Antimicrobial stewardship at transition of care from hospital to community. Infect Control Hosp Epidemiol. 2012;33(4):401–4.
86. Shah A, Petrak R, Fliegelman R, Shrestha N, Allison G, Zurlo J, et al. Infectious diseases specialty intervention is associated with better outcomes among privately insured individuals receiving outpatient parenteral antimicrobial therapy. Clin Infect Dis. 2019;68(7):1160–5.
87. Nguyen DB, Arduino MJ, Patel PR. Hemodialysis-associated infections. In: Chronic kidney disease, dialysis, and transplantation; 2019. p. 389–410.e8. https://doi.org/10.1016/B978-0-323-52978-5.00025-2.
88. Dalrymple LS, Go AS. Epidemiology of acute infections among patients with chronic kidney disease. Clin J Am Soc Nephrol. 2008;3(5):1487–93. https://doi.org/10.2215/CJN.01290308.
89. Pop-Vicas A, Strom J, Stanley K, D'Agata EM. Multidrug-resistant gram-negative bacteria among patients who require chronic hemodialysis. Clin J Am Soc Nephrol. 2008;3(3):752–8.
90. DiazGranados CA, Zimmer SM, Klein M, Jernigan JA. Comparison of mortality associated with vancomycin-resistant and vancomycin-susceptible enterococcal bloodstream infections: a meta-analysis. Clin Infect Dis. 2005;41(3):327–33.
91. Sheth H, Bernardini J, Burr R, Lee S, Miller RG, Shields M, et al. Clostridium difficile infections in outpatient dialysis cohort. Infect Control Hosp Epidemiol. 2010;31(1):89–91.
92. Klevens RM, Edwards JR, Andrus ML, et al. Dialysis surveillance report: national healthcare safety network (NHSN)-data summary for 2006. Semin Dial. 2008;21(1):24–8.
93. George A, Tokars JI, Clutterbuck EJ, Bamford KB, Pusey C, Holmes AH. Reducing dialysis associated bacteraemia, and recommendations for surveillance in the United Kingdom: prospective study. BMJ. 2006;332(7555):1435.
94. Snyder GM, Patel PR, Kallen AJ, Strom JA, Tucker JK, D'Agata EM. Antimicrobial use in outpatient hemodialysis units. Infect Control Hosp Epidemiol. 2013;34(4):349–57.
95. Snyder GM, Patel PR, Kallen AJ, Strom JA, Tucker JK, D'Agata EM. Factors associated with the receipt of antimicrobials among chronic hemodialysis patients. Am J Infect Control. 2016;44(11):1269–74.
96. Malani PN, Heung M. Optimizing antimicrobial use in hemodialysis: time to take a hard look in the mirror. Infect Control Hosp Epidemiol. 2013;34(4):358–60.
97. Hui K, Nalder M, Buising K, et al. Patterns of use and appropriateness of antibiotics prescribed to patients receiving haemodialysis: an observational study. BMC Nephrol. 2017;18(1):1–9.
98. D'Agata EMC, Tran D, Bautista J, Shemin D, Grima D. Clinical and economic benefits of antimicrobial stewardship programs in hemodialysis facilities: a decision analytic model. Clin J Am Soc Nephrol. 2018;13(9):1389–97.
99. D'Agata EMC, Lindberg CC, Lindberg CM, et al. The positive effects of an antimicrobial stewardship program targeting outpatient hemodialysis facilities. Infect Control Hosp Epidemiol. 2018;39(12):1400–5.
100. Dantes R, Mu Y, Hicks LA, Cohen J, Bamberg W, Beldavs ZG, et al. Association between outpatient antibiotic prescribing practices and community-associated *Clostridium difficile* infection. Open Forum Infect Dis. 2015;2(3):ofv113.
101. Shehab N, Lovegrove MC, Geller AI, Rose KO, Weidle NJ, Budnitz DS. US emergency department visits for outpatient adverse drug events, 2013-2014. JAMA. 2016;316(20):2115–25.
102. Slayton RB, Toth D, Lee BY, et al. Vital signs: estimated effects of a coordinated approach for action to reduce antibiotic-resistant infections in health care facilities – United States. MMWR Morb Mortal Wkly Rep. 2015;64(30):826–31.

Antibiotic Stewardship in the Emergency Department

3

Kellie J. Goodlet, Michael D. Nailor, and Larissa S. May

Introduction

Emergency department (ED) visits represent over 50% of all medical care in the United States (U.S.) [1], with nearly 140 million visits nationwide in 2015 alone [2]. Infection consistently ranks as one of the top reasons for ED presentation, and antibiotics are prescribed or administered in over 15% of all visits [2]. Although the establishment of antibiotic stewardship programs in the ED has historically lagged behind inpatient stewardship, the ED represents a unique practice setting for stewardship, whereby interventions have the potential to markedly impact both inpatient and ambulatory antibiotic prescribing. In 2013, a call to action for ED antibiotic stewardship was published in the Annals of Emergency Medicine [3], and in 2017 the Society of Infectious Diseases Pharmacists independently advocated for ED stewardship with increased training and utilization of ED pharmacists as stewardship champions [4]. Further, the American College of Emergency Physicians passed a resolution in 2018 affirming the importance of antibiotic stewardship and

K. J. Goodlet (✉)
Department of Pharmacy Practice, Midwestern University College of Pharmacy, Glendale, AZ, USA
e-mail: kgoodlet@midwestern.edu

M. D. Nailor
Department of Pharmacy Services, St. Joseph's Hospital and Medical Center, Phoenix, AZ, USA
e-mail: michael.nailor@dignityhealth.org

L. S. May
Department of Emergency Medicine, University of California Davis, Sacramento, CA, USA
e-mail: lsmay@ucdavis.edu

S. Doron, M. Campion (eds.), *Antimicrobial Stewardship in Non-Traditional Settings*, https://doi.org/10.1007/978-3-031-21710-4_3

committing to the provision of educational resources and stewardship toolkits to ED providers [5]. Thus, the ED is increasingly recognized as a key area for endorsing antibiotic stewardship principles.

Patients presenting to the ED represent a diverse population, encompassing a wide range of ages, comorbidities, and infection severity. ED clinicians serve as front-line providers to the majority of hospital inpatients, with over 80% of all admissions first receiving ED care [6]. In light of accelerating bacterial resistance, it has been increasingly recognized that efforts to promote inpatient antibiotic stewardship must begin within the ED, given that antibiotic decisions are frequently first made in this setting. ED clinicians historically have had limited patient information at their disposal relative to downstream providers, and the fast-paced setting, overcrowding, and time constraints often necessitate quick decision-making. Rapid diagnostics (Table 3.1) are beginning to change this paradigm, allowing ED clinicians to more quickly and accurately diagnose illness and expedite clinical decisions to improve patient care. Further, antibiotic therapy decisions made in the ED have been shown to substantially influence inpatient antibiotic prescribing practices. Even when ED antibiotic use can be judged inappropriate based on guideline criteria, the majority of the time inpatient providers will choose to continue an inappropriate ED regimen rather than discontinuing or modifying the antibiotic therapy [7]. Failure to reevaluate the need for antibiotics initiated in the ED can thus lead to overuse of antibiotics in hospitalized patients, making it particularly important to promote appropriate antibiotic use in the ED. Shift scheduling and high rates of clinician turnover also represent challenges, as they limit the durability of educational initiatives alone; thus, multi-modal stewardship interventions will likely be needed.

While previously neglected, the ED and urgent care settings represent the greatest opportunity for impacting patient outcomes through antibiotic stewardship. Approximately 30 million antibiotic prescriptions are written from the ED each year, ranking among the top 5 most prescribed medication classes [2]. Unfortunately, an estimated 60% of these prescriptions are guideline-discordant, with 40% prescribed without a clear clinical indication for antibiotic therapy at all [8]. Additionally, broader-spectrum antibiotics are being prescribed at increasing rates [9], including for patients at low risk for resistant organisms. This overprescription has immediate and direct risks to patient safety; for every 1000 outpatient antibiotic prescriptions, one patient will re-present to the ED due to an adverse effect. In

Table 3.1 Rapid diagnostic tests with potential to aid emergency department stewardship

Infection	Test
Upper respiratory tract infection	• Rapid viral testing (particularly for influenza) • Rapid strep test (pharyngitis)
Pneumonia	• Rapid viral testing • Procalcitonin • MRSA nasal PCR
Urinary tract infection	• *None*
Skin and soft tissue infection	• Rapid MRSA assay

aggregate, antibiotics were implicated in nearly 20% of all ED visits for drug-related adverse events [10]. Given these directly observable harms, ED clinicians are potentially the most poised to recognize unnecessary antibiotic use as a serious issue. Indeed, in a 2012–2013 survey of 150 ED providers, over three-quarters believed antibiotics are overused, though 50% admitted to overusing themselves [11]. Potential barriers to stewardship expressed by the ED providers in this survey included concerns regarding lower patient satisfaction if antibiotics are not prescribed, inadequate access to follow-up care, and lack of pathogen and susceptibility data during the ED visit [11].

Stewardship Priorities

To maximize impact, ED antibiotic stewardship programs should direct initiatives toward the most common infectious disease states encountered in the ED—upper respiratory tract infection, pneumonia, urinary tract infection, and skin and soft tissue infection. These four infection types comprise over 70% of all infection-related ED visits [9].

High-level priorities for ED stewardship include the following:

- Antibiotic therapy should not be prescribed when not clinically indicated. For example, minimal or no antibiotic use should be directed toward treatment of viral respiratory infections or asymptomatic bacteriuria. Completely avoiding inappropriate antibiotic use is likely to have the most profound impact on reducing total antibiotic utilization and decreasing subsequent resistance development and direct patient harms.
- Second, antibiotic therapy, when indicated, should be directed toward the likely etiologic organisms causing the infection. Use of broad-spectrum antibiotics, particularly those with activity against *Pseudomonas aeruginosa* or methicillin-resistant *Staphylococcus aureus* (MRSA), should be avoided for patients at low risk for drug-resistant organisms to help limit antibiotic resistance development and propensity for clonal spread. Examples include limiting fluoroquinolone use for community-acquired uncomplicated cystitis, reducing the use of gram-negative or anaerobic agents for skin and soft tissue infections, and avoiding use of broad-spectrum antibiotics such as piperacillin-tazobactam for patients with pneumonia or intra-abdominal infection who are being admitted to the hospital from the community with low risk for infection by drug-resistant bacteria.
- Finally, ED clinicians should also, when feasible, set downstream providers up for success with respect to antibiotic stewardship. Traditional cultures, although of limited utility within the ED setting, support inpatient stewardship efforts to de-escalate or discontinue antibiotics by providing objective data in support of a recommendation (e.g., negative culture or growth of susceptible pathogen). Yield of clinical cultures often declines rapidly if not taken before antibiotics are given; therefore, ED clinicians should attempt to obtain appropriate cultures prior to antibiotic administration whenever possible. Conversely, inappropriate cultures

can impede stewardship efforts when contaminating or colonizing organisms are present and prompt inappropriate antibiotic prescribing. Approximately 1–5% of blood cultures become contaminated with commensal skin flora, therefore their obtainment is discouraged for low-yield infections such as cellulitis. Similarly, indiscriminate ordering of urine cultures leads to unnecessary treatment of asymptomatic bacteriuria, which has been linked to drug resistance [12].

Additional infection-specific considerations are discussed in the following sections.

Respiratory Tract Infection

Respiratory tract infections (RTIs) include sinusitis, otitis media, pharyngitis, bronchitis, and other nonspecific upper respiratory infections, such as the common cold. Pneumonia is addressed separately in the next section. Although almost always viral in nature, upper RTI and bronchitis represent the most common indications for ambulatory antibiotic prescribing [13]. Despite decreasing antibiotic prescription rates in primary care settings, antibiotic utilization for RTI within the ED has remained high, and prescriptions for broad-spectrum antibiotics have increased significantly [14, 15]. Although rates vary widely from institution to institution, collectively 40–50% of all upper RTI visits will result in an antibiotic prescription [9, 16]. In a review of prescribing patterns at a large integrated health system, less than one-third of all sinusitis encounters resulted in guideline-consistent care [17], and the Healthcare Effectiveness Data and Information Set (HEDIS) reported that nearly three-quarters of all bronchitis patients were prescribed antibiotics against national quality standards in 2017 [18]; in a 2018 study of two large hospital EDs, the majority (77%) of these antibiotics were broad-spectrum [19]. Therefore, there is a significant need for effective antibiotic stewardship measures.

Diagnosis

Multiplex PCR respiratory viral panels (e.g., FilmArray, Verigene, Prodesse) have been proposed as a means of improving antibiotic prescribing practices for RTI by rapidly identifying non-bacterial causes of patients' symptoms. Example viruses which may be identified via these panels include respiratory syncytial virus, influenza, parainfluenza, metapneumovirus, adenovirus, rhinovirus, and coronaviruses. Stewardship guidelines by the Infectious Diseases Society of America (IDSA) and the Society for Healthcare Epidemiology of America (SHEA) recommend the use of rapid viral testing for respiratory pathogens to help eliminate unnecessary antibiotic use, though this is classified as a weak recommendation due to limited and inconsistent clinical data [20]. Meta-analyses have confirmed the diagnostic accuracy of these tests [21], which in real-world settings have improved diagnostic yield by nearly fourfold [22]. However, rapid respiratory panels have not consistently resulted in reductions in antibiotic use, with the exception of for patients with a positive influenza result [23]. In two studies evaluating their impact, despite 37% and 52% of patients testing positive for a viral pathogen, there were no differences in antibiotic prescription rates for these patients compared to pre-intervention rates or compared to patients negative for the tested viruses [22, 24]. A statistically significant difference was noted for duration of antibiotic therapy, but numerically this difference was less than a half-day [22]. By contrast, positive rapid molecular tests for influenza have been successfully linked to reductions in antibiotic prescription rates and increased receipt of appropriate antiviral therapy [23–25]. It has been postulated that rapid tests for influenza might be more successful in achieving decreases in antibiotic use compared to rapid tests for other viruses because of the availability of an antimicrobial therapeutic option, i.e. antiviral therapy for influenza, that can be prescribed in lieu of antibiotic therapy and may represent a more palatable treatment option to the provider than supportive care alone. While still a promising tool, multiplex panels are likely inadequate on their own for influencing antibiotic prescribing without supporting interventions or stewardship champions to encourage clinician behavioral change.

Treatment

Upper RTI treatment optimization should focus on minimizing unnecessary and inappropriate antibiotic use. While education-based initiatives have demonstrated modest effectiveness in some studies [26, 27], others resulted in no improvement [28, 29]. Similarly, computer decision support systems and clinical pathways incorporating guideline recommendations resulted in no or modest (<15%) decreases in antibiotic prescriptions [30–32], with the exception of for pharyngitis where more substantial increases in treatment appropriateness were noted [33, 34]. This emphasizes the importance of multi-modal stewardship interventions, such as those also incorporating provider feedback [35, 36] and clinician peer-comparison data [37] to increase the likelihood of achieving sustainable improvements. Interventions that leverage behavioral economics have been successful in preserving physician autonomy while improving guideline-concordant management. As an additional resource, the Centers for Disease Control and Prevention (CDC) provides Core Elements of hospital and outpatient antibiotic stewardship as guidance for establishing a robust stewardship program [38, 39].

Further work is needed to engage patients, given that patient expectations remain the greatest barrier to avoidance of antibiotics for viral RTI. One method which has demonstrated promise [27, 40, 41] is the provision of patient information sheets to educate on the lack of utility of antibiotics for treatment of RTI. Patient satisfaction is a common concern among ED clinicians, and it is a prevailing belief that patients are more likely to be dissatisfied with their care when antibiotics are not prescribed. In a study of ED physicians' prescribing practices, physicians were more likely to prescribe antibiotics to patients with upper RTIs who they believed expected them, although they correctly identified only about 1 in 4 of those patients [42]. Further, the study found that patient satisfaction was best attributed not to the receipt of antibiotics but rather to whether or not the patients felt that the physician had a good understanding of their illness. Therefore, it may be inferred that providers who confidently recommend against antibiotics are likely to still be rated favorably by patients. Provision of patient information sheets outlining conditions for which antibiotics are unnecessary may also increase provider confidence in withholding antibiotic therapy when it is unwarranted [43].

Another method that has demonstrated substantial reductions in rates of antibiotic use is adopting a "watch and wait" or "delayed" antibiotic prescription strategy, whereby patients are either sent home without antibiotics but told to contact their providers after a certain time period if their symptoms have not improved or patients are provided with an antibiotic prescription but with the advice to delay filling the prescription. This strategy is endorsed by the American Academy of Pediatrics as an option for children older than 6 months with mild to moderate acute otitis media [44]. Similarly, although the current IDSA rhinosinusitis guidelines recommend adoption of more stringent clinical criteria to distinguish bacterial from non-bacterial causes of sinusitis as the preferred strategy to reduce unnecessary antibiotic use rather than withholding or delaying empiric antibiotic therapy, they do acknowledge that some patients with non-severe symptoms may be reasonably observed without antibiotic treatment for 3 days [45]. In a large pooled analysis of multiple types of RTI, delayed antibiotic prescription achieved lower rates of antibiotic use compared to immediate antibiotics (31% versus 93%) with no increase in complications or decrease in patient satisfaction [46]. Ultimately, ED clinicians are encouraged to engage their RTI patients in shared decision-making discussions that address both the lack of likely benefit and potential harms of antibiotic therapy. One suggested script that accurately captures the heart of this issue is the following: "For your infection, there is an ~1 in 4000 chance that an antibiotic will prevent a serious complication, a 5–25% chance that it will cause diarrhea, and an ~1 in 1000 chance that you will require a visit to the emergency department because of a bad reaction to the antibiotic" [47].

Pneumonia

Among the most common ED infections, pneumonia arguably has the most potential for morbidity and mortality. Approximately 20% of lower RTIs result in hospitalization, and cumulatively these infections rank within the top 10 causes of death in the U.S. [9]. Although pneumonia stewardship has historically focused on optimizing empiric antibiotic therapy, the advent of new biomarkers and rapid diagnostics may, as with upper RTI, allow for better identification of non-bacterial causes of respiratory symptoms and facilitate more rapid discontinuation of antibiotic therapy.

Diagnosis

Among admitted patients with an initial pneumonia diagnosis, <15% will have an identified bacterial pathogen [48]. Although some of this can likely be attributed to the nonculturable or fastidious nature of some of the common bacteria implicated in pneumonia (e.g., the atypicals, *Streptococcus pneumoniae*), viral causes of pneumonia, not treatable by antibiotics, are also likely represented. Additionally, noninfectious "pneumonia mimics," including congestive heart failure (CHF) and chronic obstructive pulmonary disease (COPD) exacerbations, may present with nonspecific radiographic findings and respiratory symptoms (e.g., cough, sputum production, dyspnea, tachypnea) that cannot be reliably differentiated from pneumonia. Thus, being able to distinguish between bacterial and non-bacterial causes of lower respiratory symptoms would be of significant value in this setting.

Procalcitonin is an acute phase reactant which may assist with antibiotic prescribing decisions for lower RTIs by signifying the presence and severity of a bacterial infection. Clinical procalcitonin testing was approved by the U.S. Food and Drug Administration (FDA) in 2017 after already being utilized by many European countries decades prior. Procalcitonin is more specific than either C-reactive protein or erythrocyte sedimentation rate, as it selectively rises in response to systemic bacterial infections. Other favorable properties include its predictable time dynamics, increasing within 6 h upon stimulation and peaking at 6–8 h, and that procalcitonin quantitatively correlates with disease severity [49]. Initial procalcitonin levels have been used to assess the need for antibiotic initiation, while serial procalcitonin concentrations have been used in the inpatient setting to indicate a sufficient duration of therapy, facilitating timely antibiotic discontinuation. Multiple randomized controlled trials [50–52] as well as "real-world" studies [53, 54] have demonstrated consistent and significant reductions in antibiotic therapy duration with procalcitonin-guided therapy, without negatively impacting clinical outcomes for pneumonia. In a large meta-analysis of over 6700 patients across 26 trials, a 35% decrease in antibiotic consumption was observed (5.7 versus 8.1 days, $P < 0.001$), and additionally 30-day mortality was lower in the procalcitonin therapy group (adjusted OR 0.83, $P = 0.037$) [55]. This observed decrease in mortality could be attributable to confounding by severity, whereby clinicians may be more likely to discontinue antibiotics in response to a low procalcitonin level for less severely ill patients. However, potentially supporting the legitimacy of these findings, patients receiving procalcitonin-guided antibiotic therapy had a lower risk of antibiotic-related adverse effects (16.3% versus 22.1%, $P < 0.001$), and additionally CHF patients, in a previous subgroup analysis, were shown to have lower overall rates of adverse events when procalcitonin was used [56]. Intravenous antibiotics represent an insidious source of salt and fluid volume among hospitalized patients, leading to potentially harmful effects when given inappropriately to patients experiencing CHF exacerbation. The use of procalcitonin to rule out bacterial infection in these patients may allow for earlier initiation of effective therapies (e.g., diuresis).

However, in a recent U.S. randomized trial of approximately 1600 patients presenting to the ED, implementation of procalcitonin for lower RTI failed to produce a significant reduction in antibiotic days compared to usual care [57]. The control group in this study already had a relatively short mean treatment duration of 4 days, limiting the potential effect; in prior studies, control courses were longer (mean 8 days). Protocol adherence was also low; 92% of patients should not have received antibiotics based on their procalcitonin values, however 59% of these patients did. It is possible the clinicians had already decided to prescribe the antibiotic and thus the test result did not fit into clinical workflow decision support. Finally, less than 50% of patients were admitted, meaning that these patients did not have follow-up procalcitonin values to potentially reduce therapy duration.

It has also been questioned whether the sensitivity of procalcitonin for bacterial infection is adequate to justify withholding antibiotic therapy. Although not specific to pneumonia, in a study of over 300 bacteremic inpatients, over one-third of these patients had a below-threshold initial procalcitonin value, including 19% of patients admitted to an intensive care unit [58]. Also concerning, use of procalcitonin-guided antibiotic therapy for severe acute exacerbation of COPD failed to demonstrate non-inferiority to standard care with respect to 3-month mortality in a multicenter randomized trial [59]. Ultimately, use of procalcitonin is not recommended as a means of withholding empiric antibiotic therapy within the 2019 community-acquired pneumonia guidelines [60]; however, use may be considered as a supplemental stewardship tool, such as to help with shortening duration of therapy in the inpatient setting or to support a viral pneumonia diagnosis.

Utilization of real-time PCR panels for respiratory viruses and atypical bacterial pathogens also holds promise as a potential stewardship intervention by supporting discontinuation or streamlining of antibiotic therapy. However, as with upper RTI, current real-world data have been disappointing, with such tests successfully increasing diagnostic yield but failing to result in reductions in antibiotic use due to prescribers not acting upon the test's results [61]. Thus, stewardship teams should exercise caution before purchasing these testing panels, which in the aforementioned analysis increased treatment and diagnostic costs by over $300 per patient, and have a plan in place for education, implementation, and follow-up. ED pharmacists (Table 3.2) may play a role by identifying patients who may benefit from testing and following up on the test's results.

Table 3.2 Potential stewardship roles of emergency department pharmacists

Patient-level • Assist with allergy clarification and interpretation • Review prior patient antibiotic and culture history • Interpret diagnostic test results • Guide empiric antibiotic selection • Optimize antibiotic dosing and duration (outpatients) • Facilitate timely antibiotic administration • Manage potential drug–drug interactions • Facilitate transitions of care
Provider-level • Promote stewardship initiatives and institutional guidelines • Provide drug information and educational resources • Enforce formulary antibiotic restriction criteria • Provide clinician education and active feedback
Systems-level • Lead culture review and follow-up programs, including antibiotic modification/de-escalation for discharged patients • Collaborate with antibiotic stewardship team on ED-directed initiatives • Adapt national guidelines to the local context and setting via the creation of infection-specific clinical pathways • Aid in the development of emergency department-specific antibiograms • Review data on institutional antibiotic use and guidance adherence and assess intervention effectiveness

Treatment

Common bacterial pathogens among patients presenting with pneumonia from the community include *S. pneumoniae*, *Haemophilus influenzae*, and *Moraxella catarrhalis*, as well as the "atypical" organisms—*Mycoplasma pneumoniae*, *Chlamydophila pneumoniae*, and *Legionella pneumophila*. Previously, patients were classified as having either "community-acquired pneumonia (CAP)" or "healthcare-associated pneumonia (HCAP)" on the basis that patients in the latter group—which included all nursing home and dialysis patients, as well as patients with recent hospitalizations—would be at increased risk for pneumonia caused by drug-resistant bacteria including *P. aeruginosa* and MRSA and therefore require broader empiric antibiotic therapy. However, since the publication of the 2005 HCAP guidelines (now retired) [62], it has been increasingly recognized that the HCAP definition may not accurately identify patients harboring resistant pathogens [63]. In one analysis of HCAP patients, patients receiving guideline-concordant antibiotic regimens (active against CAP pathogens plus *P. aeruginosa* and MRSA) did not have higher clinical cure rates than HCAP patients who empirically received CAP regimens (with no or limited *P. aeruginosa* and MRSA activity) [64]. Explaining this finding, multiple subsequent studies have demonstrated that the independent predictive value of any one HCAP risk factor for drug-resistant pathogens is low, and not all factors confer equal risk. Instead, multiple risk factors are likely needed to have a >10% likelihood of a resistant pathogen [65–67]. Several alternative scoring systems have been proposed to better identify at-risk patients [65, 68, 69]; however, currently no single model has demonstrated universally high sensitivity and specificity [70]. Thus, institution-specific ED pneumonia guidelines, tailored to the local epidemiology, are more likely to be effective in this setting for accurately risk-stratifying patients with pneumonia and minimizing unnecessary broad-spectrum antibiotic use. Local guidelines, in combination with education and prospective audit and feedback, have been effective at reducing length of stay and increasing the proportion of patients receiving concordant therapy [71–73]. Additionally, ED providers are strongly encouraged to obtain respiratory cultures despite their low yield, as finalized negative cultures have been shown to be an appropriate indication for de-escalation of broad-spectrum antibiotic therapy [74]. Similarly, the MRSA nasal swab PCR assay has a >99% negative predictive value for MRSA pneumonia among HCAP patients and can additionally support de-escalation efforts by downstream providers [75].

Additional successful stewardship interventions include the utilization of computerized decision support systems to optimize CAP prescribing, which have demonstrated more substantial improvements in concordant therapy rates than academic detailing alone [76], reductions in length of stay [77], and lower observed than expected mortality [78]. Implementation of a review and feedback initiative also resulted in modest improvements in initial empiric antibiotic selection in the ED [79], and simply preloading appropriate CAP antibiotics in automated dispensing systems resulted in substantial improvements in empiric antibiotic selection in combination with the development of a CAP treatment algorithm [80]. These and other user-friendly interventions that may be readily incorporated into the busy ED workflow are more likely to be utilized and result in practice change than more lengthy and complex guideline documents [81].

Finally, in the outpatient setting, ED clinicians are encouraged to prescribe shorter duration (5–7 day) therapy and avoid macrolide monotherapy (Table 3.3).

Table 3.3 Recommended antibiotic regimens and durations for common outpatient bacterial infections

Infection	Organisms	Recommended ABX	ABX to Avoid	Duration
Community-acquired pneumonia—No DRSP risk factors[a]	*S. pneumoniae*, *H. influenzae*, *M. catarrhalis*, the "atypical" respiratory bacteria, viruses	Doxycycline (preferred) Azithromycin	β-lactam monotherapy	5 days
Community-acquired pneumonia—(+) DRSP risk factors[a]	*S. pneumoniae*, *H. influenzae*, *M. catarrhalis*, the "atypical" respiratory bacteria, viruses	Cefpodoxime + [doxycycline or macrolide] Amoxicillin/clavulanate + [doxycycline or macrolide] Respiratory fluoroquinolone	Macrolide or doxycycline monotherapy, β-lactam monotherapy	5 days
Uncomplicated cystitis	*E. coli*, *S. saprophyticus*, *Klebsiella* spp., *Proteus* spp.	Nitrofurantoin TMP-SMX[b]	Fluoroquinolone	5 days 3 days
Uncomplicated pyelonephritis	*E. coli*, *Klebsiella* spp., *Proteus* spp.	Ciprofloxacin or levofloxacin [+/− parenteral ceftriaxone or aminoglycoside[c]] TMP-SMX	Nitrofurantonin (cystitis only)	7 days
Cellulitis	*S. pyogenes*, other streptococci, *S. aureus*	Cephalexin Doxycycline Dicloxacillin	Dual therapy, Gram-negative agents (if non-bite-associated)	5 days
Cutaneous abscess	*S. aureus* (MSSA/MRSA), *S. pyogenes*, other streptococci	Consider none if uncomplicated and adequate incision and drainage. Alternatively, addition of: TMP-SMX Doxycycline Clindamycin	Gram-negative agents	N/A, or 5 days

Abbreviations: ABX antibiotics, DRSP drug-resistant *Streptococcus pneumoniae*, N/A not applicable, TMP-SMX trimethoprim-sulfamethoxazole

[a]Risk factors include antibiotic therapy in the past 3 months, chronic heart, lung, liver, or renal disease, diabetes mellitus, alcoholism, malignancy, asplenia, or immunocompromise

[b]If community resistance <20%

[c]If community resistance to fluoroquinolone ≥10%

S. pneumoniae remains the single most common bacterial cause of CAP [82], and high resistance rates limit the utility of macrolides for bacterial CAP when used alone, particularly when risk factors for *S. pneumoniae* drug resistance are present [83]. Combination therapy with an appropriate β-lactam antibiotic or alternate therapy with a respiratory fluoroquinolone for high-risk patients, or doxycycline monotherapy for low-risk patients, may be more efficacious options. Doxycycline additionally has the advantage of being associated with a lower risk of *Clostridioides difficile* infection [84].

Urinary Tract Infections

Although urinary tract infection (UTI), which encompasses conditions such as cystitis and pyelonephritis, represents an appropriate indication for antibiotic use in the ED, misdiagnosis and inappropriate antibiotic selection lead to national guideline non-adherence rates of 60% and higher.

Diagnosis

Improving diagnosis represents the greatest potential UTI stewardship opportunity in the ED, in particular distinguishing true UTI from asymptomatic bacteriuria (ASB). Approximately 5% of all young, healthy, sexually active women and 11–16% of elderly men and women from the community will have ASB [85, 86]. ASB prevalence increases to up to 50% among adults residing in long-term care facilities, while virtually all patients (100%) with indwelling urinary catheters will become colonized within 2 weeks of placement [87]. Even when catheterized patients are excluded, inpatients with ASB outnumber inpatients with UTI by over sixfold [88]. With few exceptions (e.g., pregnancy and immediately before an invasive urologic procedure), patients with ASB should not receive antibiotic therapy [86]. In addition to the broader stewardship benefits of withholding antibiotic therapy for non-infected patients, ASB has been shown to have a protective role in preventing symptomatic UTI. In a 12-month follow-up study, ASB-treated women showed higher rates of UTI recurrence than women who did not receive treatment for ASB (47% versus 13%, $P < 0.0001$), and their UTIs were caused by progressively more resistant organisms [12]. Despite these consequences, ASB treatment is common, with 40–50% of these patients receiving antibiotics [88, 89].

Interventions to limit ASB treatment are predominantly directed at reducing the ordering of urinalysis and urine cultures for non-symptomatic patients. Although urinalysis is useful in supporting a suspected UTI diagnosis in the setting of clinical symptoms (e.g., dysuria, frequency, urgency), no component of the urinalysis is diagnostic of UTI on its own, and neither colony counts nor the presence of nitrite or pyuria distinguish ASB from UTI [90]. However, in many institutions, urinalysis is ordered indiscriminately for patients entering the ED, with a high proportion (~40%) auto-reflexing to culture [91]. Although inclusion of urinalysis within default order sets can be blamed for much of this overutilization, in one study where urinalysis was performed at the discretion of ED or general medicine physicians, over half (62%) of all patients still underwent urinalysis; 84% of these patients lacked urinary symptoms [92]. Other studies assessing the appropriateness of urine cultures have similarly found that the majority (58–68%) of urine cultures are ordered without a guideline-based clinical indication [93, 94]. Abnormal urinalysis results have been shown to be a trigger for guideline-discordant prescribing behavior by introducing cognitive biases that favor a UTI diagnosis even in the absence of symptoms or other guideline-based indications for antibiotic therapy. For example, older patient age, pyuria, and colonization of the urine by gram-negative rods have all been associated with an increased likelihood of antibiotic prescription despite not being indications for treatment [90].

A simple intervention to reduce overall antibiotic prescribing for ASB is to apply more stringent criteria for urinalysis reflex to culture. In one study, the investigators modified their criteria such that urine reflexed to culture only if pyuria ≥10/hpf was present *plus* the presence of blood, protein ≥30 mg/dL, leukocyte esterase, nitrite, or cloudiness. A significant decrease in reflexed cultures was observed, resulting in an estimated 702 courses of inappropriate antibiotics saved per year, despite no change in the rate of ASB treatment pre- and post-intervention [91]. Going further, another institution completely suppressed urine culture results for non-catheterized patients; clinicians had to call the microbiology lab if they wished to know the culture results. Post-intervention, clinicians inquired about the suppressed culture results for only 14% of cases, and treatment of ASB decreased from 48% to 12% ($P = 0.002$). For the four patients with UTI during the study period, there was no intervention-associated delay in antibiotic therapy; all patients received empiric antibiotics based on the presence of urinary symptoms [88]. Additional strategies include promotion of good culturing technique to reduce contamination and laboratory rejection of contaminated specimens (e.g., >5 epithelial cells per low power field) [95].

Special patient populations pose unique diagnostic challenges and may require additional educational efforts by the stewardship team to help clinicians avoid unneeded antibiotic prescriptions. Elderly patients, particularly those residing in skilled nursing facilities, who present to the ED with altered mental status may have a decreased ability to endorse urinary symptoms, leading some clinicians to diagnose UTI and initiate antibiotic therapy on the basis of altered mentation alone. In one study, confusion was the number one reason urine cultures were ordered for otherwise asymptomatic patients [96]. However, given the high prevalence of ASB (15–50% [86]) as well as dementia (63% [97]) in the skilled nursing population, diagnosing UTI on the basis of bacteriuria plus cognitive impairment alone would appear to result in a large portion of these patients receiving unnecessary therapy. Additionally, although earlier studies reported an associated between falls and UTI among the elderly, the best prospective evidence to date demonstrated no association unless UTI symptoms were also present [98]. Indeed, changes in mental function and other nonspecific signs such as falls, malaise, and changes in voiding patterns were found to be poorly predictive even of bacteriuria with pyuria [97].

There are also numerous known alternative sources of altered mental status in the elderly, including dehydration, hypoxia, and adverse drug reactions. Thus, for elderly patients presenting with confusion alone not in septic shock, it has been recommended to observe these patients off antibiotics for 24–48 h [99]. Often, altered mental status will resolve with fluids or oxygen supplementation alone. Implementation of symptom-based minimum criteria for empiric UTI antibiotic initiation in long-term care residents (Fig. 3.1), [100] has also been shown to be successful at decreasing antibiotic use compared to standard care without negatively impacting clinical outcomes [101]. This may also be a preferred approach for other patient populations less able to mount systemic signs of infection such as leukocytosis (e.g., immunosuppression).

ED clinicians should also be trained to recognize patient populations at high risk for ASB, such as dialysis patients and patients with spinal cord injury. Although these specific groups cannot be as readily diagnosed on a symptom basis, their high rate of bacterial colonization means that UTI should be considered a diagnosis of exclusion. Among patients with indwelling catheters in particular, although dysuria is not assessable, patients should still be evaluated for

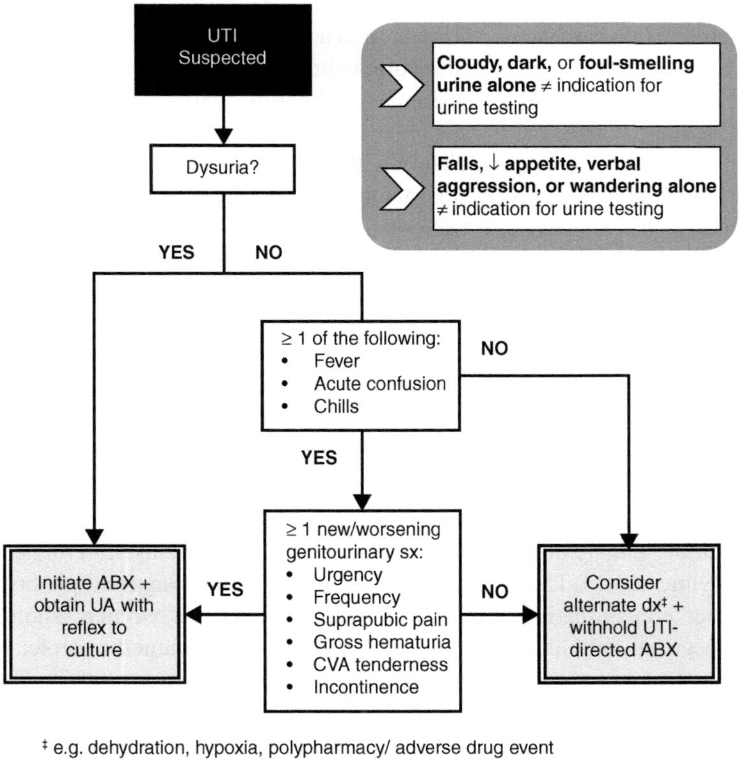

Fig. 3.1 Approach to workup of urinary tract infection diagnosis for patients presenting from long-term care facilities. Created as part of an institutional urinary tract infection guideline at St. Joseph's Hospital and Medical Center, Phoenix, AZ. *ABX* antibiotics, *UA* urinalysis, *sx* symptoms, *dx* diagnosis, *CVA* costovertebral angle

other infection signs and symptoms such as leukocytosis and suprapubic pain or tenderness before initiating UTI treatment, given virtually all patients will have bacteriuria [87]. Fever in isolation should be interpreted cautiously; in one study, more than one-third of catheterized febrile patients had a clearly identifiable non-urinary cause of fever [102]. Similar to the work-up and evaluation of elderly patients with altered mental status, clinicians may consider adopting a "wait-and-see" approach when considering a diagnosis of catheter-associated UTI (CA-UTI). In a prospective cohort study, appropriate early empiric treatment of patients diagnosed with CA-UTI (~80% febrile) was not associated with reductions in 30-day mortality, suggesting that antibiotic therapy may be safely deferred until a more thorough patient work-up may be performed in the inpatient setting [103]. From a CA-UTI prevention perspective, ED clinicians should also strive to promote multidisciplinary efforts to reduce unnecessary urinary catheter use. Inappropriate indications for catheterization include nursing convenience and urine output monitoring; this unnecessary use is estimated to represent up to 65% of all catheter use, placing patients at risk for preventable CA-UTI [104, 105]. Catheter use should be in accordance with CDC guidelines and kept in place only as long as is medically necessary. Potentially preferred alternatives to chronic indwelling urethral catheters to reduce CA-UTI risk include intermittent catheterization and external (condom) catheters [104], which in a randomized controlled trial demonstrated a significant decrease in the composite outcome of bacteriuria, symptomatic UTI, and death when compared to indwelling catheter use among males without dementia [106].

Even if antibiotics are appropriately withheld for ASB in the ED, there is potential for downstream providers to inappropriately initiate antibiotic therapy in response to urinalysis results or bacteriuria. Thus, to aid inpatient stewardship, it is recommended that ED clinicians explicitly document the lack of urinary symptoms and the deliberate decision to withhold antibiotics (e.g. "Asymptomatic bacteriuria ((-)dysuria, (-)frequency, (-)urgency. No antibiotic therapy indicated.").

Treatment

UTIs are caused predominantly by ascending host bowel flora. Enterobacterales, in particular *Escherichia coli*, represent the vast majority of isolates, with other organisms including *Staphylococcus saprophyticus* and uncommonly *Enterococcus faecalis*. Polymicrobial UTIs are rare, and urine cultures showing multiple organisms may reflect contamination. For uncomplicated cystitis in women, short-course narrow-spectrum antibiotic therapy with trimethoprim-sulfamethoxazole, nitrofurantoin, or fosfomycin is recommended for outpatient therapy [107]. However, broad-spectrum antibiotics are being prescribed at increasing rates in the ED. Fluoroquinolones continue to be the most commonly prescribed class of antibiotic for UTI, despite increasing resistance and a 2016 warning issued by the FDA stating that the potential for serious adverse effects outweighs the benefit for patients with uncomplicated cystitis [9]. The majority (60%) of UTI antibiotic prescriptions are non-guideline adherent, largely driven by inappropriate fluoroquinolone use [96].

A variety of strategies have been employed to reduce fluoroquinolone use for cystitis. The most restrictive, removal of fluoroquinolones from the ED formulary, was successful in combination with the provision of a UTI antibiotic suggestion list in significantly reducing ciprofloxacin use at one Norwegian institution [108]. More commonly, computer decision support tools such as electronic order sets for UTI have been developed to guide clinician antibiotic selection [109, 110]. Limitations include the potential for low provider uptake [109] if the order set is not readily accessible or easily bypassed. Potentially more effective is the combination of computer decision support tools with audit and feedback; one facility employing this strategy saw a near doubling of guideline-concordant antibiotic selection (44–82%), including a 31% reduction in fluoroquinolone use. Greater results were observed with the addition of audit and feedback than with order set implementation alone [111]. However, UTI misdiagnosis rates remained high (>40%), underscoring the need for stewardship interventions aimed at reducing inappropriate antibiotic initiation in order to have the greatest impact on reducing antibiotic use.

The creation of ED-specific antibiograms has also been used as a tool to optimize UTI prescribing in the ED. Antibiograms are a valuable aid for guiding empiric antibiotic selection, and clinicians are less likely to prescribe non-susceptible antibiotics when local susceptibilities are disseminated [112]. However, hospital antibiograms overestimate community resistance and tend to be biased toward high-risk patients or those presenting with treatment failure. Narrower-spectrum, guideline-recommended antibiotics may appear less effective than broader agents, discouraging clinician use. Antibiograms specific to the ED thus may result in improved susceptibilities for guideline-recommended agents and promote increased prescribing. Going further, some facilities have also demonstrated success with providing ED antibiograms stratified by patient disposition, such as ED-only versus inpatient [113–115].

When inpatient admission is not indicated, UTI patients are discharged home on antibiotics prior to knowing definitive culture and susceptibility results. Thus, structured culture follow-up allows for clinicians to identify discordant antibiotic therapy and provide timely definitive therapy when clinically indicated. Effective culture follow-up results in an increased proportion of patients receiving effective therapy and has been associated with reductions in ED revisits and 30-day hospitalization [116, 117]. Culture follow-up also provides an opportunity to discontinue unnecessary antibiotic therapy, such as for patients with negative urine cultures [89, 118]. ED pharmacists (Table 3.2) are particularly well-suited to take ownership of culture follow-up programs, as they generally have increased flexibility in their shift. When culture review was performed by pharmacists, there were fewer missed interventions [119, 120], shortened time to prescriber notification of culture results [118, 121], and increased appropriateness of definitive therapy [122] compared to culture review by nurses. Pharmacist culture review was also equivalent to or better than ED physician review [117].

Finally, for outpatient UTI therapy, the stewardship team should advocate for the prescribing of the shortest effective antibiotic duration (Table 3.3). For example,

7 days of fluoroquinolone therapy for pyelonephritis has been shown to be noninferior to 14 days, even among patients with bacteremia [123].

Skin and Soft Tissue Infections

Skin and soft tissue infections (SSTIs) are a common and growing reason for ED presentation. From 1993 to 2005, the number of ED visits for SSTIs nearly tripled [124], and SSTI ranks among the top 5 presenting diagnoses requiring subsequent hospital admission among 18–64 year olds [125]. However, as with UTI, misdiagnosis is common [126] and antibiotic coverage for SSTI is frequently broader than necessary [127]. Thus, SSTI represents a prime area for targeted stewardship interventions.

Diagnosis

Misdiagnosis of cellulitis leads to an estimated 50,000–130,000 unnecessary hospitalizations and up to $515 million in excess costs to the healthcare system annually [126]. Noninfectious conditions that can mimic the symptoms of lower extremity cellulitis (e.g., erythema, swelling, pain), often grouped together under the category of "pseudocellulitis," include venous stasis or contact dermatitis, CHF-associated edema, deep vein thrombosis, gout, and hematoma, among others. Pseudocellulitis represents 30–40% of all cellulitis diagnoses and is a common cause of inappropriate antibiotic therapy in the ED [126, 128]. It is thus recommended for stewardship teams to educate ED clinicians on these cellulitis mimics and encourage clinicians to first rule out edema-causing conditions before assigning a cellulitis diagnosis, especially for patients without systemic signs of infection. In particular, bilateral cellulitis is extremely rare and almost always representative of CHF edema or stasis dermatitis. As with CHF exacerbations mimicking pneumonia, administration of intravenous antibiotics for CHF edema pseudocellulitis may harm patients by increasing salt and fluid overload. Infectious diseases specialist follow-up has been shown to be effective in the outpatient setting at reducing the days of unnecessary antibiotic therapy due to SSTI misdiagnosis [128].

Treatment

Source control is a vital component of SSTI treatment when abscess is present. Uncomplicated abscesses that are incised and drained may be appropriately treated without antibiotic therapy, and avoidance of antibiotics for these patients is a "Choosing Wisely" recommendation by the American College of Emergency Physicians [129]. This is supported by meta-analytic data that determined that systemic antibiotic therapy did not increase the percentage of patients with complete abscess resolution when compared to incision and drainage alone [130, 131] and did not reduce the risk for abscess recurrence [131]. In assessing ED prescribing practices, antibiotics are often overprescribed for discharged patients with drained abscesses, representing an area of opportunity for stewardship [132]. It should be acknowledged that in two recent placebo-controlled trials, statistically significant improvements in cure rates were seen when 7–10 days of trimethoprim-sulfamethoxazole or clindamycin was given to patients with drained uncomplicated skin abscess [133, 134]. However, limitations of these trials should be noted. In one trial [134], a substantial (>20%) portion of patients had significant (>75 cm^2) surrounding erythema; thus, the improvements observed with antibiotic therapy may potentially be attributed to appropriate treatment of concomitant cellulitis. In the other trial [133], "skin infection at new body site" was included as a criterion for treatment failure, biasing the results toward antibiotic therapy due to skin decolonization rather than enhanced abscess resolution. When this criterion is removed, no statistical difference in cure rates was observed between the antibiotic and incision and drainage-only groups [135]. It has also been noted that even assuming true benefit, the number needed to treat with antibiotics to achieve one additional cure is relatively high at 7–14. Even erring on the side of greatest potential antibiotic effect, this would still imply that antibiotic therapy will be unnecessary for ≥70% of patients [136]. Notably, no cases of sepsis or infection-attributed mortality were reported among the >2000 included patients. Conversely, adverse effects with antibiotic therapy were relatively common (up to 22% with clindamycin, number needed to harm = 10). Thus from a risk–benefit perspective, withholding of antibiotic therapy for patients with drained uncomplicated abscess remains a reasonable front-line approach.

Although most nonpurulent SSTIs are not routinely cultured in clinical practice, based on clinical trial data it is generally accepted that most cases of cellulitis are caused by *Streptococcus pyogenes* and other *Streptococcus spp*, as well as *S. aureus* [137, 138]. Purulent SSTIs, including abscesses, are most commonly caused by *S. aureus*, including MRSA, although streptococci may be isolated from abscesses as well. Thus, antibiotic treatment of SSTI is directed toward streptococci and staphylococci, with the decision to administer antibiotics active against MRSA dependent upon clinical presentation (e.g., presence of carbuncle or complicated abscess) and patient specific risk factors (e.g., injection drug use or known colonization). "Dual coverage" for SSTI, whereby cephalexin is added in addition to trimethoprim-sulfamethoxazole therapy for patients with MRSA risk factors based on in vitro data suggesting reduced trimethoprim-sulfamethoxazole activity against some *Streptococcus spp*, is not recommended given clinical data demonstrating the adequacy of trimethoprim-sulfamethoxazole monotherapy [139].

A substantial increase in MRSA-targeted empiric SSTI regimens has been observed in a relatively short period of time, nearly doubling from 38% of all SSTI regimen in 2005 to 74% in 2010 [124]. Correspondingly, antibiotic management of SSTI has shifted from predominant use of cephalosporins to sulfonamides [9]. This is despite multiple modern randomized trials failing to show higher rates of clinical resolution of nonpurulent cellulitis when an MRSA-active antibiotic is given with cephalexin therapy compared to cephalexin therapy alone [140, 141]. In addition to reflecting potential overuse, this is a stewardship concern given that the most commonly prescribed oral antibiotics for MRSA, sulfonamides and clindamycin, also have the highest rates of adverse effects leading to ED representation per prescription [10]. One proposed intervention to improve empiric gram-positive therapy is the utilization of rapid molecular testing for *S. aureus*. For example, the GeneXpert PCR (Cepheid, Sunnyvale CA) can detect and distinguish MSSA and MRSA from a wound swab within 60 min [142]. With a high (>97%) negative predictive value for MRSA, theoretically the test could reduce unnecessary use of MRSA-active antibiotics. However, while turnaround time was shown to be adequate to ensure receipt of results prior to patient discharge [143], real-world implementation results have been mixed. Although associated with more targeted antibiotic therapy in some studies of lower-risk patients not requiring inpatient admission [144], others have failed to demonstrate statistically significant reductions in overprescription of MRSA-active antibiotics due to limited acceptance of results by clinicians [143, 145]. In one study of hospitalized patients admitted from the ED, 65% of patients without MRSA still received MRSA-active antibiotics despite active pharmacist notification of the results with associated provider feedback; this was not significantly different from the rate of discordant therapy at baseline [145]. This underscores that clinicians may be more hesitant to trust the results of rapid diagnostics and withhold broad-spectrum therapy for higher acuity patients. Although potentially more likely to impact antibiotic selection for outpatients, another limitation is that the test is only effective for purulent SSTIs. With uncomplicated drained abscesses often not requiring antibiotic therapy at all, this limits test utility. More effectively, a multi-modal stewardship intervention involving educational presentations by a designated physician champion, implementation of an electronic order set based on IDSA SSTI guidelines, dissemination of an ED-specific *S. aureus* wound isolate antibiogram, and regular audit and feedback including peer-comparison successfully decreased clindamycin use from 58% to 28% with a corresponding increase in cephalexin use from 22% to 42% [146].

Gram-negative bacteria and anaerobes play a limited role in uncomplicated SSTIs; however, routine coverage is generally indicated for infections more likely to be polymicrobial, which include diabetic foot ulcers, bite wounds, necrotizing fasciitis, and select cases of SSTI associated with trauma. Adherence to bite wound guidelines tends to be suboptimal (62% in one study [147]), and implementation of electronic order sets may help steer ED clinicians away from ineffective options such as cephalexin or macrolides. However, more commonly, empiric gram-negative coverage for SSTI is broader than indicated. In one study assessing the epidemiology and treatment of SSTIs in the ED, 30% of patients received antibiotics with expanded gram-negative activity (approximately half of which were active against *P. aeruginosa*); however, gram-negative organisms represented only 17% of all organisms ultimately isolated. Of these, there was only a single isolate of *P. aeruginosa* [148]. In another study focusing on patients requiring hospitalization, the overprescription of broad-spectrum antibiotics was even more pronounced, with extended-spectrum gram-negative and anaerobic antibiotics being prescribed to 61–80% and 73–83% of patients, respectively. Nearly all (97%) of these patients ultimately grew *S. aureus* or streptococci [149]. Although surveillance data of SSTIs among hospitalized patients indicate up to a 10% rate of *P. aeruginosa*, clinicians should recognize that this likely represents an overestimation of the true prevalence, as these rates are biased toward readily culturable pyogenic wound infections, with severe or recurrent infections more likely to be cultured than uncomplicated cases [150]. Even with diabetic foot ulcers, where use of gram-negative and anaerobic antibiotics is warranted, empiric use of antibiotics with *P. aeruginosa* activity should be discouraged for hemodynamically stable patients, as regimens without *P. aeruginosa* activity have demonstrated equal effectiveness, even in cases where *P. aeruginosa* was later isolated [151]. Institution-specific SSTI treatment guidelines may assist with guiding appropriate empiric therapy.

Increasing interest has also been given to transitions of care interventions, particularly with respect to averting inpatient admission for SSTI. In one analysis, the mean total cost of care for a SSTI inpatient was $13,313 versus only $413 for an outpatient [148]. Long-acting lipoglycopeptides, oritavancin and dalbavancin, require only a single dose for SSTI treatment, which has been proposed to facilitate a rapid transition to outpatient care, including for patients who would otherwise be admitted to hospital care [152]. Both drugs have demonstrated potential cost savings in cost-minimization models projecting reduced admissions [153, 154] and may be considered a preferable alternative to multi-day oral therapy by SSTI patients [155]. However, data demonstrating real-world feasibility and savings are lacking. When compared to conventional outpatient parenteral antibiotic therapy in a retrospective analysis of SSTI patients, dalbavancin was not cost-effective (mean $4561 infectious diseases-related cost of care versus $1668 for conventional IV therapy) except when compared to daptomycin ($5218) [156]. Additional research is needed to determine if long-acting glycopeptides will reduce hospitalization and thus prove to be cost-effective in certain situations. In general, the need for receipt of IV antibiotics has fallen off intensity of care criteria needed to justify hospitalization for insurance purposes and is more sensitive to the instability of the patient. As such, it is unclear if ED providers will infuse long-acting glycopeptides into mildly unstable patients and clear them for discharge and what outcomes will be for these patients. Alternatively, long-acting glycopeptides may be relegated for patients who would already be eligible for oral therapy and be given based on patient preference or presumed lack of adherence to oral therapy. However, routine use in the ED setting cannot be recommended at this time.

Finally, stewardship teams should promote judicious ordering of cultures for SSTI. Most cases of cellulitis and uncomplicated abscess do not require cultures, as the impact on definitive therapy is minimal provided that appropriate empiric therapy was given. Swabbing of skin wounds should be avoided, as these are likely to grow commensal skin organisms and thus have limited clinical utility. Obtaining blood cultures is also not recommend for immunocompetent SSTI patients not in septic shock due to low (<5%) yield and the risk for false positives due to contamination by skin flora [157]. Additional "low hanging fruit" for antibiotics in SSTI management in ED patients may be reducing duration of antibiotics for those patients discharged with SSTI (Table 3.3).

Conclusions

The ED represents an untapped landscape for antibiotic stewardship programs to profoundly impact both inpatient and outpatient prescribing practices. Given the unique barriers to stewardship that are present in ED settings, multi-modal stewardship interventions are needed to enact enduring change. High-value interventions are directed toward reducing inappropriate use of antibiotics and promoting the most appropriate narrow-spectrum agents when antibiotic therapy is indicated. Appropriate culture collection and communication of diagnostic uncertainty is encouraged to assist inpatient stewardship efforts.

Conflict of Interest Statement Dr. Nailor has served as a consultant for Shionogi, AbbVie, and Merck. Dr. May has received grant funding from Inflammatix, Roche Diagnostics, and Roche Molecular Systems. She has received speaking honoraria from ThermoFisher and Inflammatix and has served as a consultant or advisory board member for BioRad, Cepheid, Abbott, Roche Diagnostics, and Roche Molecular Systems.

References

1. Marcozzi D, Carr B, Liferidge A, Baehr N, Browne B. Trends in the contribution of emergency departments to the provision of hospital-associated health care in the USA. Int J Health Serv. 2018;48:267–88. https://doi.org/10.1177/0020731417734498.
2. National Hospital Ambulatory Medical Care Survey emergency department summary tables emergency department summary tables; 2018. https://www.cdc.gov/nchs/data/nhamcs/web_tables/2015_ed_web_tables.pdf. Accessed 29 Jan 2019.
3. May L, Cosgrove S, L'Archeveque M, Talan DA, Payne P, Jordan J, et al. A call to action for antimicrobial stewardship in the emergency department: approaches and strategies. Ann Emerg Med. 2013;62:69–77. https://doi.org/10.1016/j.annemergmed.2012.09.002.
4. Dilworth T, Delgado G, Dionne B, McDaneld P, Khan S, Wrenn R, et al. SIDP position statement on the role of pharmacists in antibiotic stewardship in the emergency department. In: SIDP political advocacy; 2017. https://www.sidp.org/resources/Documents/Position%20Statements/ED%20Pharmacists%20in%20stewardship%209-20.pdf. Accessed 14 Feb 2019.
5. American College of Emergency Physicians. 2018 resolutions compendium; 2018. http://www.rsc.org/dose/title_of_subordinate_document. Accessed 28 Mar 2019.
6. Gonzalez Morganti K, Bauhoff S, Blanchard JC, Abir M, Iyer N, Smith AC, et al. The evolving roles of emergency departments in the United States. Rand Health Q. 2013;3(2):3. https://www.rand.org/content/dam/rand/pubs/research_reports/RR200/RR280/RAND_RR280.pdf. Accessed 29 Jan 2019.
7. Kiyatkin D, Bessman E, McKenzie R. Impact of antibiotic choices made in the emergency department on appropriateness of antibiotic treatment of urinary tract infections in hospitalized patients. J Hosp Med. 2016;11:181–4. https://doi.org/10.1002/jhm.2508.
8. Grenet J, Davido B, Bouchand F, Sivadon-Tardy V, Beauchet A, Tritz T, et al. Evaluating antibiotic therapies prescribed to adult patients in the emergency department. Med Mal Infect. 2016;46:207–14. https://doi.org/10.1016/j.medmal.2016.04.006.
9. Mohareb AM, Dugas AF, Hsieh YH. Changing epidemiology and management of infectious diseases in US EDs. Am J Emerg Med. 2016;34:1059–65. https://doi.org/10.1016/j.ajem.2016.03.024.
10. Shehab N, Patel PR, Srinivasan A, Budnitz DS. Emergency department visits for antibiotic-associated adverse events. Clin Infect Dis. 2008;47:735–43. https://doi.org/10.1086/591126.

11. May L, Gudger G, Armstrong P, Brooks G, Hinds P, Bhat R, et al. Multisite exploration of clinical decision making for antibiotic use by emergency medicine providers using quantitative and qualitative methods. Infect Control Hosp Epidemiol. 2014;35:1114–25. https://doi.org/10.1086/677637.
12. Cai T, Mazzoli S, Mondaini N, Meacci F, Nesi G, D'Elia C, et al. The role of asymptomatic bacteriuria in young women with recurrent urinary tract infections: to treat or not to treat? Clin Infect Dis. 2012;55:771–7. https://doi.org/10.1093/cid/cis534.
13. Fleming-Dutra KE, Hersh AL, Shapiro DJ, Bartoces M, Enns EA, File TM Jr, et al. Prevalence of inappropriate antibiotic prescriptions among US ambulatory care visits, 2010-2011. JAMA. 2016;315:1864–73. https://doi.org/10.1001/jama.2016.4151.
14. Donnelly JP, Baddley JW, Wang HE. Antibiotic utilization for acute respiratory tract infections in U.S. emergency departments. Antimicrob Agents Chemother. 2014;58:1451–7. https://doi.org/10.1128/AAC.02039-13.
15. Grijalva CG, Nuorti JP, Griffin MR. Antibiotic prescription rates for acute respiratory tract infections in US ambulatory settings. JAMA. 2009;302:758–66. https://doi.org/10.1001/jama.2009.1163.
16. Gonzales R, Camargo CA Jr, MacKenzie T, Kersey AS, Maselli J, Levin SK, et al. Antibiotic treatment of acute respiratory infections in acute care settings. Acad Emerg Med. 2006;13:288–94.
17. Sharp AL, Klau MH, Keschner D, Macy E, Tang T, Shen E, et al. Low-value care for acute sinusitis encounters: who's choosing wisely? Am J Manag Care. 2015;21:479–85.
18. National Committee for Quality Assurance. Avoidance of antibiotic treatment in adults with acute bronchitis: percentage of adults 18 to 64 years of age with a diagnosis of acute bronchitis who were not dispensed an antibiotic prescription; 2017. https://www.ncqa.org/hedis/measures/avoidance-of-antibiotic-treatment-in-adults-with-acute-bronchitis/. Accessed 29 Jan 2019.
19. Kroening-Roche JC, Soroudi A, Castillo EM, Vilke GM. Antibiotic and bronchodilator prescribing for acute bronchitis in the emergency department. J Emerg Med. 2012;43:221–7. https://doi.org/10.1016/j.jemermed.2011.06.143.
20. Barlam TF, Cosgrove SE, Abbo LM, MacDougall C, Schuetz AN, Septimus EJ, et al. Implementing an antibiotic stewardship program: guidelines by the Infectious Diseases Society of America and the Society for Healthcare Epidemiology of America. Clin Infect Dis. 2016;62:e51–77. https://doi.org/10.1093/cid/ciw118.
21. Huang HS, Tsai CL, Chang J, Hsu TC, Lin S, Lee CC. Multiplex PCR system for the rapid diagnosis of respiratory virus infection: systematic review and meta-analysis. Clin Microbiol Infect. 2018;24:1055–63. https://doi.org/10.1016/j.cmi.2017.11.018.
22. Rogers BB, Shankar P, Jerris RC, Kotzbauer D, Anderson EJ, Watson JR, et al. Impact of a rapid respiratory panel test on patient outcomes. Arch Pathol Lab Med. 2015;139:636–41. https://doi.org/10.5858/arpa.2014-0257-OA.
23. Semret M, Schiller I, Jardin BA, Frenette C, Loo VG, Papenburg J, et al. Multiplex respiratory virus testing for antimicrobial stewardship: a prospective assessment of antimicrobial use and clinical outcomes among hospitalized adults. J Infect Dis. 2017;216:936–44. https://doi.org/10.1093/infdis/jix288.
24. Green DA, Hitoaliaj L, Kotansky B, Campbell SM, Peaper DR. Clinical utility of on-demand multiplex respiratory pathogen testing among adult outpatients. J Clin Microbiol. 2016;54:2950–5.
25. Dugas A, Rothman R. Clinical utility of rapid molecular-based influenza testing. Open Forum Infect Dis. 2015;2(Suppl 1):102. https://doi.org/10.1093/ofid/ofv131.22.
26. Gkentzi D, Ramachandran R, Day E, Anguvaa L, Borg M, Siva N, et al. Antibiotic prescribing in the paediatric emergency department and the impact of education. J Paediatr Child Health. 2014;50:932–3. https://doi.org/10.1111/jpc.12753.
27. Småbrekke L, Berild D, Giaever A, Myrbakk T, Fuskevåg A, Ericson JU, et al. Educational intervention for parents and healthcare providers leads to reduced antibiotic use in acute otitis media. Scand J Infect Dis. 2002;34:657–9.

28. Célind J, Södermark L, Hjalmarson O. Adherence to treatment guidelines for acute otitis media in children. The necessity of an effective strategy of guideline implementation. Int J Pediatr Otorhinolaryngol. 2014;78:1128–32. https://doi.org/10.1016/j.ijporl.2014.04.029.
29. Palma S, Rosafio C, Del Giovane C, Patianna VD, Lucaccioni L, Genovese E, et al. The impact of the Italian guidelines on antibiotic prescription practices for acute otitis media in a paediatric emergency setting. Ital J Pediatr. 2015;41:37. https://doi.org/10.1186/s13052-015-0144-4.
30. Gonzales R, Anderer T, McCulloch CE, Maselli JH, Bloom FJ Jr, Graf TR, et al. A cluster randomized trial of decision support strategies for reducing antibiotic use in acute bronchitis. JAMA Intern Med. 2013;173:267–73. https://doi.org/10.1001/jamainternmed.2013.1589.
31. Linder JA, Schnipper JL, Tsurikova R, Yu T, Volk LA, Melnikas AJ, et al. Documentation-based clinical decision support to improve antibiotic prescribing for acute respiratory infections in primary care: a cluster randomised controlled trial. Inform Prim Care. 2009;17:231–40.
32. Samore MH, Bateman K, Alder SC, Hannah E, Donnelly S, Stoddard GJ, et al. Clinical decision support and appropriateness of antimicrobial prescribing: a randomized trial. JAMA. 2005;294:2305–14.
33. Diaz MC, Symons N, Ramundo ML, Christopher NC. Effect of a standardized pharyngitis treatment protocol on use of antibiotics in a pediatric emergency department. Arch Pediatr Adolesc Med. 2004;158:977–81.
34. Woolley SL, Bernstein JM, Davidson JA, Smith DR. Sore throat in adults--does the introduction of a clinical scoring system improve the management of these patients in a secondary care setting? J Laryngol Otol. 2005;119:550–5.
35. Metlay JP, Camargo CA Jr, MacKenzie T, McCulloch C, Maselli J, Levin SK, et al. Cluster-randomized trial to improve antibiotic use for adults with acute respiratory infections treated in emergency departments. Ann Emerg Med. 2007;50:221–30.
36. Ouldali N, Bellêttre X, Milcent K, Guedj R, de Pontual L, Cojocaru B, et al. Impact of implementing national guidelines on antibiotic prescriptions for acute respiratory tract infections in pediatric emergency departments: an interrupted time series analysis. Clin Infect Dis. 2017;65:1469–76. https://doi.org/10.1093/cid/cix590.
37. Meeker D, Linder JA, Fox CR, Friedberg MW, Persell SD, Goldstein NJ, et al. Effect of behavioral interventions on inappropriate antibiotic prescribing among primary care practices: a randomized clinical trial. JAMA. 2016;315:562–70. https://doi.org/10.1001/jama.2016.0275.
38. Centers for Disease Control and Prevention. Core elements of hospital antibiotic stewardship programs. Atlanta: US Department of Health and Human Services, CDC; 2014. http://www.cdc.gov/getsmart/healthcare/implementation/core-elements.html. Accessed 30 Jan 2019.
39. Sanchez GV, Fleming-Dutra KE, Roberts RM, Hicks LA. Core elements of outpatient antibiotic stewardship. MMWR Recomm Rep. 2016;65:1–12.
40. Sanchez X, Tello B, Jimbo R, Douce RW, Remmen R, Henriquez R. Implementation of NICE 69 clinical guideline reduced inappropriate antibiotic prescriptions for upper respiratory tract infections, in an emergency department in Quito, Ecuador. Eur J Gen Pract. 2013;19:43.
41. Sun C, Nitzberg M, Herzberg DL, Bhat R. Antibiotic stewardship of acute respiratory infections in the emergency department. Acad Emerg Med. 2016;23:S26.
42. Ong S, Nakase J, Moran GJ, Karras DJ, Kuehnert MJ, Talan DA. Antibiotic use for emergency department patients with upper respiratory infections: prescribing practices, patient expectations, and patient satisfaction. Ann Emerg Med. 2007;50:213–20.
43. Centers for Disease Control and Prevention. Be antibiotics aware partner toolkit; 2018. https://www.cdc.gov/antibiotic-use/week/toolkit.html. Accessed 14 Feb 2019.
44. Lieberthal AS, Carroll AE, Chonmaitree T, Ganiats TG, Hoberman A, Jackson MA, et al. The diagnosis and management of acute otitis media. Pediatrics. 2013;131:e964–99. https://doi.org/10.1542/peds.2012-3488.
45. Chow AW, Benninger MS, Brook I, Brozek JL, Goldstein EJ, Hicks LA, et al. IDSA clinical practice guideline for acute bacterial rhinosinusitis in children and adults. Clin Infect Dis. 2012;54:e72–e112. https://doi.org/10.1093/cid/cir1043.

46. Spurling GK, Del Mar CB, Dooley L, Foxlee R, Farley R. Delayed antibiotic prescriptions for respiratory infections. Cochrane Database Syst Rev. 2017 Sep;7(9):CD004417. https://doi.org/10.1002/14651858.CD004417.pub5.
47. Linder JA. Editorial commentary: antibiotics for treatment of acute respiratory tract infections: decreasing benefit, increasing risk, and the irrelevance of antimicrobial resistance. Clin Infect Dis. 2008;47:744–6. https://doi.org/10.1086/591149.
48. Jain S, Self WH, Wunderink RG, Fakhran S, Balk R, Bramley AM, et al. Community-acquired pneumonia requiring hospitalization among U.S. adults. N Engl J Med. 2015;373:415–27. https://doi.org/10.1056/NEJMoa1500245.
49. Meisner M. Pathobiochemistry and clinical use of procalcitonin. Clin Chim Acta. 2002;323:17–29.
50. Christ-Crain M, Stolz D, Bingisser R, Mueller C, Leuppi J, Battegay M, et al. Procalcitonin guidance significantly reduces antibiotic duration in community-acquired pneumonia: the 'ProCAP' study. Crit Care. 2005;9(Suppl 1):P166.
51. Schuetz P, Christ-Crain M, Thomann R, Falconnier C, Wolbers M, Widmer I, et al. Effect of procalcitonin-based guidelines vs standard guidelines on antibiotic use in lower respiratory tract infections: the ProHOSP randomized controlled trial. JAMA. 2009;302:1059–66. https://doi.org/10.1001/jama.2009.1297.
52. Tang J, Long W, Yan L, Zhang Y, Xie J, Lu G, et al. Procalcitonin guided antibiotic therapy of acute exacerbations of asthma: a randomized controlled trial. BMC Infect Dis. 2013;13:596. https://doi.org/10.1186/1471-2334-13-596.
53. Albrich WC, Dusemund F, Bucher B, Meyer S, Thomann R, Kühn F, et al. Effectiveness and safety of procalcitonin-guided antibiotic therapy in lower respiratory tract infections in "real life": an international, multicenter poststudy survey (ProREAL). Arch Intern Med. 2012;172:715–22. https://doi.org/10.1001/archinternmed.2012.770.
54. Stojanovic I, Schneider JE, Wei L, Hong Z, Keane C, Schuetz P. Economic evaluation of procalcitonin-guided antibiotic therapy in acute respiratory infections: a Chinese hospital system perspective. Clin Chem Lab Med. 2017;55:561–70. https://doi.org/10.1515/cclm-2016-0349.
55. Schuetz P, Wirz Y, Sager R, Christ-Crain M, Stolz D, Tamm M, et al. Procalcitonin to initiate or discontinue antibiotics in acute respiratory tract infections. Cochrane Database Syst Rev. 2017;10:CD007498. https://doi.org/10.1002/14651858.CD007498.pub3.
56. Schuetz P, Kutz A, Grolimund E, Haubitz S, Demann D, Vögeli A, et al. Excluding infection through procalcitonin testing improves outcomes of congestive heart failure patients presenting with acute respiratory symptoms: results from the randomized ProHOSP trial. Int J Cardiol. 2014;175:464–72. https://doi.org/10.1016/j.ijcard.2014.06.022.
57. Huang DT, Yealy DM, Filbin MR, Brown AM, Chang CH, Doi Y, et al. Procalcitonin-guided use of antibiotics for lower respiratory tract infection. N Engl J Med. 2018;379:236–49. https://doi.org/10.1056/NEJMoa1802670.
58. Goodlet KJ, Cameron EA, Nailor MD. Low sensitivity of procalcitonin for bacteremia at an academic medical center: a cautionary tale for antimicrobial stewardship. Open Forum Infect Dis. 2020;7(4):ofaa096. https://doi.org/10.1093/ofid/ofaa096.
59. Daubin C, Valette X, Thiollière F, Mira JP, Hazera P, Annane D, Labbe V, Floccard B, Fournel F, Terzi N, Du Cheyron D, Parienti JJ, BPCTrea Study Group. Procalcitonin algorithm to guide initial antibiotic therapy in acute exacerbations of COPD admitted to the ICU: a randomized multicenter study. Intensive Care Med. 2018;44(4):428–37. https://doi.org/10.1007/s00134-018-5141-9.
60. Metlay JP, Waterer GW, Long AC, Anzueto A, Brozek J, Crothers K, Cooley LA, Dean NC, Fine MJ, Flanders SA, Griffin MR, Metersky ML, Musher DM, Restrepo MI, Whitney CG. Diagnosis and treatment of adults with community-acquired pneumonia. An official clinical practice guideline of the American Thoracic Society and Infectious Diseases Society of America. Am J Respir Crit Care Med. 2019;200(7):e45–67. https://doi.org/10.1164/rccm.201908-1581ST.

61. Oosterheert JJ, van Loon AM, Schuurman R, Hoepelman AI, Hak E, Thijsen S, et al. Impact of rapid detection of viral and atypical bacterial pathogens by real-time polymerase chain reaction for patients with lower respiratory tract infection. Clin Infect Dis. 2005;41:1438–44.
62. American Thoracic Society. Infectious Diseases Society of America. Guidelines for the management of adults with hospital-acquired, ventilator-associated, and healthcare-associated pneumonia. Am J Respir Crit Care Med. 2005;171:388–416.
63. Chalmers JD, Rother C, Salih W, Ewig S. Healthcare-associated pneumonia does not accurately identify potentially resistant pathogens: a systematic review and meta-analysis. Clin Infect Dis. 2014;58:330–9. https://doi.org/10.1093/cid/cit734.
64. Chen JI, Slater LN, Kurdgelashvili G, Husain KO, Gentry CA. Outcomes of health care-associated pneumonia empirically treated with guideline-concordant regimens versus community-acquired pneumonia guideline-concordant regimens for patients admitted to acute care wards from home. Ann Pharmacother. 2013;47:9–19. https://doi.org/10.1345/aph.1R322.
65. Shindo Y, Ito R, Kobayashi D, Ando M, Ichikawa M, Shiraki A, et al. Risk factors for drug-resistant pathogens in community-acquired and healthcare-associated pneumonia. Am J Respir Crit Care Med. 2013;188:985–95. https://doi.org/10.1164/rccm.201301-0079OC.
66. Shorr AF, Zilberberg MD, Micek ST, Kollef MH. Prediction of infection due to antibiotic-resistant bacteria by select risk factors for health care-associated pneumonia. Arch Intern Med. 2008;168:2205–10. https://doi.org/10.1001/archinte.168.20.2205.
67. Webb BJ, Dascomb K, Stenehjem E, Dean N. Predicting risk of drug-resistant organisms in pneumonia: moving beyond the HCAP model. Respir Med. 2015;109:1–10. https://doi.org/10.1016/j.rmed.2014.10.017.
68. Prina E, Ranzani OT, Polverino E, Cillóniz C, Ferrer M, Fernandez L, et al. Risk factors associated with potentially antibiotic-resistant pathogens in community-acquired pneumonia. Ann Am Thorac Soc. 2015;12:153–60. https://doi.org/10.1513/AnnalsATS.201407-305OC.
69. Webb BJ, Dascomb K, Stenehjem E, Vikram HR, Agrwal N, Sakata K, et al. Derivation and multicenter validation of the drug resistance in pneumonia clinical prediction score. Antimicrob Agents Chemother. 2016;60:2652–63. https://doi.org/10.1128/AAC.03071-15.
70. Self WH, Wunderink RG, Williams DJ, Barrett TW, Baughman AH, Grijalva CG. Comparison of clinical prediction models for resistant bacteria in community-onset pneumonia. Acad Emerg Med. 2015;22:730–40. https://doi.org/10.1111/acem.12672.
71. Almatar M, Peterson GM, Thompson A, McKenzie D, Anderson T, Zaidi ST. Clinical pathway and monthly feedback improve adherence to antibiotic guideline recommendations for community-acquired pneumonia. PLoS One. 2016;11:e0159467. https://doi.org/10.1371/journal.pone.0159467.
72. Julián-Jiménez A, Palomo de los Reyes MJ, Laín-Terés N. Is it possible to improve the management of community acquired pneumonia in hospital emergency departments? Arch Bronconeumol. 2010;46:448–9. https://doi.org/10.1016/j.arbres.2010.03.001.
73. Julián-Jiménez A, Palomo de los Reyes MJ, Parejo Miguez R, Laín-Terés N, Cuena-Boy R, Lozano-Ancín A. Improved management of community-acquired pneumonia in the emergency department. Arch Bronconeumol. 2013;49:230–40. https://doi.org/10.1016/j.arbres.2012.12.008.
74. Zilberberg MD, Shorr AF. Healthcare-associated pneumonia: the state of evidence to date. Curr Opin Pulm Med. 2011;17:142–7. https://doi.org/10.1097/MCP.0b013e328343eb33.
75. Dangerfield B, Chung A, Webb B, Seville MT. Predictive value of methicillin-resistant Staphylococcus aureus (MRSA) nasal swab PCR assay for MRSA pneumonia. Antimicrob Agents Chemother. 2014;58:859–64. https://doi.org/10.1128/AAC.01805-13.
76. Buising KL, Thursky KA, Black JF, MacGregor L, Street AC, Kennedy MP, et al. Improving antibiotic prescribing for adults with community acquired pneumonia: Does a computerised decision support system achieve more than academic detailing alone?--A time series analysis. BMC Med Inform Decis Mak. 2008;8:35. https://doi.org/10.1186/1472-6947-8-35.
77. Marrie TJ, Lau CY, Wheeler SL, Wong CJ, Vandervoort MK, Feagan BG. A controlled trial of a critical pathway for treatment of community-acquired pneumonia. JAMA. 2000;283:749–55.

78. Dean NC, Jones BE, Jones JP, Ferraro JP, Post HB, Aronsky D. Impact of an electronic clinical decision support tool for emergency department patients with pneumonia. Ann Emerg Med. 2015;66:511–20. https://doi.org/10.1016/j.annemergmed.2015.02.003.
79. McIntosh KA, Maxwell DJ, Pulver LK, Horn F, Robertson MB, Kaye KI, et al. A quality improvement initiative to improve adherence to national guidelines for empiric management of community-acquired pneumonia in emergency departments. Int J Qual Health Care. 2011;23:142–50. https://doi.org/10.1093/intqhc/mzq077.
80. Ostrowsky B, Sharma S, DeFino M, Guo Y, Shah P, McAllen S, et al. Antimicrobial stewardship and automated pharmacy technology improve antibiotic appropriateness for community-acquired pneumonia. Infect Control Hosp Epidemiol. 2013;34:566–72. https://doi.org/10.1086/670623.
81. Mistry RD, Dayan PS, Kuppermann N. The battle against antimicrobial resistance: time for the emergency department to join the fight. JAMA Pediatr. 2015;169:421–2. https://doi.org/10.1001/jamapediatrics.2015.79.
82. Mandell LA, Bartlett JG, Dowell SF, File TM Jr, Musher DM, Whitney C. Update of practice guidelines for the management of community-acquired pneumonia in immunocompetent adults. Clin Infect Dis. 2003;37:1405–33.
83. Mandell LA, Wunderink RG, Anzueto A, Bartlett JG, Campbell GD, Dean NC, et al. Infectious Diseases Society of America/American Thoracic Society consensus guidelines on the management of community-acquired pneumonia in adults. Clin Infect Dis. 2007;44(Suppl 2):S27–72.
84. Doernberg SB, Winston LG, Deck DH, Chambers HF. Does doxycycline protect against development of Clostridium difficile infection? Clin Infect Dis. 2012;55:615–20. https://doi.org/10.1093/cid/cis457.
85. Hooton TM, Scholes D, Stapleton AE, Roberts PL, Winter C, Gupta K, et al. A prospective study of asymptomatic bacteriuria in sexually active young women. N Engl J Med. 2000;343:992–7.
86. Nicolle LE, Gupta K, Bradley SF, Colgan R, DeMuri GP, Drekonja D, et al. Clinical practice guideline for the management of asymptomatic bacteriuria: 2019 update by the Infectious Diseases Society of America. Clin Infect Dis. 2019;40(5):643–54. https://doi.org/10.1093/cid/ciy1121.
87. Stark RP, Maki DG. Bacteriuria in the catheterized patient. What quantitative level of bacteriuria is relevant? N Engl J Med. 1984;311:560–4.
88. Leis JA, Rebick GW, Daneman N, Gold WL, Poutanen SM, Lo P, et al. Reducing antimicrobial therapy for asymptomatic bacteriuria among noncatheterized inpatients: a proof-of-concept study. Clin Infect Dis. 2014;58:980–3. https://doi.org/10.1093/cid/ciu010.
89. Zhang X, Rowan N, Pflugeisen BM, Alajbegovic S. Urine culture guided antibiotic interventions: a pharmacist driven antimicrobial stewardship effort in the ED. Am J Emerg Med. 2017;35:594–8. https://doi.org/10.1016/j.ajem.2016.12.036.
90. Lin E, Bhusal Y, Horwitz D, Shelburne SA 3rd, Trautner BW. Overtreatment of enterococcal bacteriuria. Arch Intern Med. 2012;172:33–8. https://doi.org/10.1001/archinternmed.2011.565.
91. Wold K, Brock J, Percival K, Rearigh L, Vocelka L, Afroze A. Assessment of reflex urine culture criteria changes and its impact on treatment of asymptomatic bacteriuria. Open Forum Infect Dis. 2017;4(Suppl 1):S346.
92. Yin P, Kiss A, Leis JA. Urinalysis orders among patients admitted to the general medicine service. JAMA Intern Med. 2015;175:1711–3. https://doi.org/10.1001/jamainternmed.2015.4036.
93. Hartley S, Valley S, Kuhn L, Washer LL, Gandhi T, Meddings J, et al. Inappropriate testing for urinary tract infection in hospitalized patients: an opportunity for improvement. Infect Control Hosp Epidemiol. 2013;34:1204–7. https://doi.org/10.1086/673449.
94. Leis JA, Gold WL, Daneman N, Shojania K, McGeer A. Downstream impact of urine cultures ordered without indication at two acute care teaching hospitals. Infect Control Hosp Epidemiol. 2013;34:1113–4. https://doi.org/10.1086/673151.

cellulitis in the emergency department: a double cohort study. Diagn Microbiol Infect Dis. 2017;87:371–5. https://doi.org/10.1016/j.diagmicrobio.2016.12.015.
129. ACEP Joins choosing wisely campaign. ACEP Now. http://www.acepnow.com/article/acep-joins-choosing-wisely-campaign/. Accessed 19 Jan 2019.
130. Fahimi J, Singh A, Frazee BW. The role of adjunctive antibiotics in the treatment of skin and soft tissue abscesses: a systematic review and meta-analysis. CJEM. 2015;17:420–32. https://doi.org/10.1017/cem.2014.52.
131. Singer AJ, Thode HC Jr. Systemic antibiotics after incision and drainage of simple abscesses: a meta-analysis. Emerg Med J. 2014;31:576–8. https://doi.org/10.1136/emermed-2013-202571.
132. Pallin DJ, Camargo CA Jr, Schuur JD. Skin infections and antibiotic stewardship: analysis of emergency department prescribing practices, 2007-2010. West J Emerg Med. 2014;15:282–9. https://doi.org/10.5811/westjem.2013.8.18040.
133. Daum RS, Miller LG, Immergluck L, Fritz S, Creech CB, Young D, et al. A placebo-controlled trial of antibiotics for smaller skin abscesses. N Engl J Med. 2017;376:2545–55. https://doi.org/10.1056/NEJMoa1607033.
134. Talan DA, Mower WR, Krishnadasan A, Abrahamian FM, Lovecchio F, Karras DJ, et al. Trimethoprim-sulfamethoxazole versus placebo for uncomplicated skin abscess. N Engl J Med. 2016;374:823–32. https://doi.org/10.1056/NEJMoa1507476.
135. Yeoh DK, Blyth CC, Bowen AC. A trial of antibiotics for smaller skin abscesses. N Engl J Med. 2017;377:e36. https://doi.org/10.1056/NEJMc1711124.
136. Pulia M, Redwood R, May L. Antimicrobial stewardship in the emergency department. Emerg Med Clin North Am. 2018;36:853–72. https://doi.org/10.1016/j.emc.2018.06.012.
137. Diekema DJ, Pfaller MA, Schmitz FJ, Smayevsky J, Bell J, Jones RN, et al. Survey of infections due to Staphylococcus species: frequency of occurrence and antimicrobial susceptibility of isolates collected in the United States, Canada, Latin America, Europe, and the Western Pacific region for the SENTRY Antimicrobial Surveillance Program, 1997-1999. Clin Infect Dis. 2001;32(Suppl 2):S114–32.
138. DiNubile MJ, Lipsky BA. Complicated infections of skin and skin structures: when the infection is more than skin deep. J Antimicrob Chemother. 2004;53(Suppl 2):ii37–50.
139. Miller LG, Daum RS, Creech CB, Young D, Downing MD, Eells SJ, et al. Clindamycin versus trimethoprim-sulfamethoxazole for uncomplicated skin infections. N Engl J Med. 2015;372:1093–103. https://doi.org/10.1056/NEJMoa1403789.
140. Moran GJ, Krishnadasan A, Mower WR, Abrahamian FM, LoVecchio F, Steele MT, et al. Effect of cephalexin plus trimethoprim-sulfamethoxazole vs cephalexin alone on clinical cure of uncomplicated cellulitis: a randomized clinical trial. JAMA. 2017;317:2088–96. https://doi.org/10.1001/jama.2017.5653.
141. Pallin DJ, Binder WD, Allen MB, Lederman M, Parmar S, Filbin MR, et al. Clinical trial: comparative effectiveness of cephalexin plus trimethoprim-sulfamethoxazole versus cephalexin alone for treatment of uncomplicated cellulitis: a randomized controlled trial. Clin Infect Dis. 2013;56:1754–62. https://doi.org/10.1093/cid/cit122.
142. Xpert MRSA/SA SSTI package brochure. Sunnyvale: Cepheid; 2010.
143. Pulia MS, Calderone M, Hansen B, Stake CE, Cichon M, Li Z, et al. Feasibility of rapid polymerase chain reaction for detection of methicillin-resistant Staphylococcus aureus colonization among emergency department patients with abscesses. Open Access Emerg Med. 2013;5:17–22. https://doi.org/10.2147/OAEM.S50226.
144. May LS, Rothman RE, Miller LG, Brooks G, Zocchi M, Zatorski C, et al. A randomized clinical trial comparing use of rapid molecular testing for Staphylococcus aureus for patients with cutaneous abscesses in the emergency department with standard of care. Infect Control Hosp Epidemiol. 2015;36:1423–30. https://doi.org/10.1017/ice.2015.202.
145. Terp S, Krishnadasan A, Bowen W, Joo J, Furoy D, Chan J, et al. Introduction of rapid methicillin-resistant Staphylococcus aureus polymerase chain reaction testing and antibiotic selection among hospitalized patients with purulent skin infections. Clin Infect Dis. 2014;58:e129–32. https://doi.org/10.1093/cid/ciu039.

146. Trajano R, Ondak S, Tancredi D, Wiedeman J, Cohen SH, Miller L, et al. Emergency department specific antimicrobial stewardship intervention reduces antibiotic duration and selection for discharged adult and pediatric patients with skin and soft-tissue infections. Open Forum Infect Dis. 2017;4(Suppl 1):S274.
147. Birdsey M, Edwards G, Abetz J, Jennings N, Mitra B. Bite wounds and antibiotic prescription among patients presenting to an Australian emergency department. Int Emerg Nurs. 2016;27:42–5. https://doi.org/10.1016/j.ienj.2015.10.007.
148. Linder KE, Nicolau DP, Nailor MD. Epidemiology, treatment, and economics of patients presenting to the emergency department for skin and soft tissue infections. Hosp Pract. 1995;2017(45):9–15. https://doi.org/10.1080/21548331.2017.1279519.
149. Jenkins TC, Sabel AL, Sarcone EE, Price CS, Mehler PS, Burman WJ. Skin and soft-tissue infections requiring hospitalization at an academic medical center: opportunities for antimicrobial stewardship. Clin Infect Dis. 2010;51:895–903. https://doi.org/10.1086/656431.
150. Moet GJ, Jones RN, Biedenbach DJ, Stilwell MG, Fritsche TR. Contemporary causes of skin and soft tissue infections in North America, Latin America, and Europe: report from the SENTRY Antimicrobial Surveillance Program (1998-2004). Diagn Microbiol Infect Dis. 2007;57:7–13.
151. Lipsky BA, Armstrong DG, Citron DM, Tice AD, Morgenstern DE, Abramson MA. Ertapenem versus piperacillin/tazobactam for diabetic foot infections (SIDESTEP): prospective, randomised, controlled, double-blinded, multicentre trial. Lancet. 2005;366:1695–703.
152. Almarzoky Abuhussain SS, Goodlet KJ, Nailor MD, Nicolau DP. Optimizing skin and skin structure infection outcomes: considerations of cost of care. Expert Rev Pharmacoecon Outcomes Res. 2018;18:235–44. https://doi.org/10.1080/14737167.2018.1450142.
153. Agarwal R, Bartsch SM, Kelly BJ, Prewitt M, Liu Y, Chen Y, et al. Newer glycopeptide antibiotics for treatment of complicated skin and soft tissue infections: systematic review, network meta-analysis and cost analysis. Clin Microbiol Infect. 2018;24:361–8. https://doi.org/10.1016/j.cmi.2017.08.028.
154. Lodise TP, Fan W, Sulham KA. Economic impact of oritavancin for the treatment of acute bacterial skin and skin structure infections in the emergency department or observation setting: cost savings associated with avoidable hospitalizations. Clin Ther. 2016;38:136–48. https://doi.org/10.1016/j.clinthera.2015.11.014.
155. Almarzoky Abuhussain SS, Burak MA, Kohman KN, Jacknin G, Tart SB, Hobbs ALV, et al. Patient preferences for treatment of acute bacterial skin and skin structure infections in the emergency department. BMC Health Serv Res. 2018;18:932. https://doi.org/10.1186/s12913-018-3751-0.
156. Turco NJ, Kane-Gill SL, Hernandez I, Oleksiuk LM, D'Amico F, Pickering AJ. A cost-minimization analysis of dalbavancin compared to conventional therapy for the outpatient treatment of acute bacterial skin and skin-structure infections. Expert Opin Pharmacother. 2018;19:319–25. https://doi.org/10.1080/14656566.2018.1442439.
157. Mills AM, Chen EH. Are blood cultures necessary in adults with cellulitis? Ann Emerg Med. 2005;45:548–9.

Antimicrobial Stewardship in Critical Access Hospitals

4

Sarah B. Green and David H. Priest

Introduction

Critical access hospitals (CAH) are small medical facilities that serve residents in rural areas that might not otherwise have access to emergency medical care. The CAH designation was created in 1997 as part of the Balanced Budget Act in order to help maintain essential medical care and ensure the financial viability of small, remote hospitals [1]. Despite their smaller size and more limited clinical scope, CAHs can have antimicrobial utilization and resistance patterns similar to larger hospitals. Optimizing antimicrobial use through antimicrobial stewardship program (ASP) initiatives has consistently been associated with reductions in adverse effects, antimicrobial-resistance rates, and *Clostridioides difficile* infections, as well as drug costs [2]. As a result, ASPs are equally important in the CAH setting. In addition, ASP implementation is now required of CAHs in order for those facilities to remain eligible for Medicare funding [3]. With funding dependent on implementation of regulatory compliant ASPs, CAHs across the country are in varying stages of ASP implementation. We discuss here the unique barriers to implementation of antimicrobial stewardship programs at critical access hospitals, as well as key strategies for success.

S. B. Green (✉)
Department of Pharmacy, Emory University Hospital, Atlanta, GA, USA
e-mail: sarah.b.green@emoryhealthcare.org

D. H. Priest
Novant Health Institute for Safety and Quality, Novant Health, Winston-Salem, NC, USA
e-mail: dhpriest@novanthealth.org

S. Doron, M. Campion (eds.), *Antimicrobial Stewardship in Non-Traditional Settings*, https://doi.org/10.1007/978-3-031-21710-4_4

Critical Access Hospital Designation

At present, there are 1351 critical access hospitals across 45 states [4]. This represents 22% of US hospitals, but only 3% of licensed hospital beds [5]. The main criteria for CAH designation include a maximum of 25 acute care beds, annual average length of stay less than 4 days, and location more than 35 miles from the closest hospital. CAHs must provide 24-h emergency services but may include additional inpatient, outpatient, surgical, and long-term care services depending on the needs of the surrounding community [4] (Table 4.1). CAH designation confers financial benefit in the form of cost-based reimbursement from Medicare. CAHs may be reimbursed at a rate of cost +1% with capital improvements eligible for inclusion. Additional funding may be received from State Medicaid programs or grants. Access to specialized education and technical resources is an additional benefit that may be utilized for ASP implementation at CAHs [1].

Table 4.1 Select characteristics of critical access hospitals [4]

Characteristics	Description
General	Participation in Medicare program
	Licensed as an acute care hospital
	Compliance with federal, state, and local laws and regulations
	No more than 25 inpatient beds
	Average annual length of stay less than 96 h
Location	Rural area or treatment as rural
	More than 35 miles from nearest hospital via primary roads
	More than 15 miles from nearest hospital via secondary or mountainous terrain
	State designation as a necessary provider prior to 2006
Staffing	At least one doctor of medicine or osteopathy
	May include a physician assistant, nurse practitioner, and/or clinical nurse specialist
	Experienced emergency care personnel within 30 or 60 min of CAH depending on location
	Nurse on duty when CAH has inpatients
Required services	24-h emergency services and associated equipment
	Basic laboratory services
Optional services	Food and nutrition
	Surgical procedures
	Long-term care services ("swing-beds")
	Inpatient psychiatric care
	Inpatient rehabilitation

Antimicrobial Utilization and Resistance

A study utilizing voluntarily reported data to the National Healthcare Safety Network (NHSN) antimicrobial use (AU) system described antimicrobial consumption across a health-system of 4 large community hospitals (>200 beds) and 15 small community hospitals including 5 CAHs [6]. Over a 3-year period, antibiotic prescribing and consumption in the small community hospitals (median 436 days of therapy (DOT)/1000 patient-days (PD), range 406–597 DOT/1000PD) was comparable to that of large community hospitals (median 509 DOT/1000PD, range 134–671 DOT/1000PD). The authors highlight that this similarity was in the setting of a significantly lower case-mix index for the small hospitals, suggesting antimicrobial overuse at these smaller facilities [7]. The highest rates of overall and broad-spectrum antimicrobial use were observed in intensive care units (ICUs) regardless of facility size. Notably, while all 4 large hospitals had active ASPs led by infectious diseases experts, none of the small community hospitals in the study had ASPs [6].

Similar results were described in a surveillance study of 185 French healthcare facilities [8]. The study sought to determine if the proportion of acute beds to total beds in a facility was a better predictor of antimicrobial consumption than hospital size. Correlation between antibiotic consumption and number of beds was low ($R^2 = 0.067$), but correlation between antibiotic consumption and percent of acute care beds was high ($R^2 = 0.574$). Small facilities, defined as less than 100 beds, accounted for 19.1% of the highest antimicrobial consumption. A comparable study of 130 US hospitals found that not only number of hospital beds, but also normalized ICU days, number of surgical procedures, and cases of pneumonia, bacteremia, and urinary tract infection were strong predictors of antimicrobial use [9].

Additional studies have confirmed antimicrobial selection and indication in small and critical access hospitals are both comparable to those in larger facilities. A 2011 point-prevalence study of 5860 patients across 183 hospitals found that 38.2% of patients in small hospitals (<150 beds) received at least 1 antimicrobial agent compared to 43.1% of patients at larger facilities (150–399 beds) and 18.8% of patients at facilities with greater than 400 beds [10]. The most targeted disease states were infections of the lower respiratory tract, urinary tract, and skin and soft tissue across all hospitals. Vancomycin, piperacillin-tazobactam, and fluoroquinolones were among the most common agents utilized. Similar results were observed in a recent study with pneumonia and urinary tract infections accounting for 40% of antimicrobial indications and vancomycin and piperacillin-tazobactam being the top 2 antimicrobials utilized [11].

Due to the perceived higher patient acuity at most large community hospitals, rates of antimicrobial resistance at these facilities may be expected to be higher compared to those at small hospitals, including CAHs. This is supported by studies that have found fewer infections by multidrug-resistant organisms (MDROs) and

Pseudomonas spp. [7] and in one study that looked at causes of hospital-acquired bloodstream infection, a higher percentage of infections due to *Escherichia coli* in smaller community hospitals [12]. However, a study utilizing microbiologic data from 300 hospitals across the USA found that small community hospitals overall have similar antimicrobial-resistance rates to large, tertiary hospitals [13]. With increasing rates of MDROs in the community and long-term care facilities, infections by these organisms can be expected in CAHs as well.

Antimicrobial Stewardship Program Requirements

The Medicare Rural Hospital Flexibility Program (Flex) is a Medicare funding program designed to promote quality improvement, benchmarking, performance enhancement, population health, and emergency medical services (EMS). In order to continue to receive funding from the Flex Program, CAHs must comply with each of the following Centers for Disease Control and Prevention (CDC) Core Elements of Antimicrobial Stewardship:

- Leadership Commitment—*dedication of necessary resources*
- Accountability—*leader responsible for outcomes*
- Pharmacy Expertise—*designated pharmacist leader*
- Action—*implementation of recommended ASP interventions*
- Tracking—*monitoring antimicrobial consumption and resistance*
- Reporting—*updates to facility staff and leadership*
- Education—*education on current prescribing recommendations*

Adapted from: CDC Core Elements [14].

These requirements were added in 2017 by the Federal Office of Rural Health Policy (FORHP) to the Medicare Beneficiary Quality Improvement Project (MBQIP) in anticipation of inclusion of ASP implementation in the final rule of the updated Medicare and Medicaid Conditions of Participation (CoP). To meet the requirements of MBQIP Patient Safety measure Activity 1.01b, CAHs must enroll in NHSN and complete the Patient Safety Component Annual Hospital Survey. The survey consists of 10 required and 10 optional questions on ASP practices at the facility in order to assess the degree of implementation of the CDC core elements (Table 4.2).

Table 4.2 Summary of MBQIP ASP survey questions [15]

Required questions	Our facility has a formal statement of support for antibiotic stewardship
	Facility leadership has demonstrated a commitment to antibiotic stewardship efforts
	Our facility has a committee responsible for antibiotic stewardship
	Our facility has a leader responsible for antibiotic stewardship outcomes
	Our facility has a policy or formal procedure for documentation of indication/duration/etc. for antibiotic orders
	Providers have access to facility- or region-specific treatment guidelines or recommendations for commonly encountered infections
	Our facility targets select diagnoses for active interventions to optimize antibiotic use
	Our stewardship team monitors select metrics [e.g. antibiotic use (DDD/1000 patient days), antibiotic expenditures, C. difficile rates, antibiotic resistance]
	Our stewardship team provides the updates/reports to leadership/staff at least annually
	Prescribers/nursing staff/pharmacists receive education on appropriate antibiotic use at least annually
Optional questions	Antibiotic stewardship activities are integrated into quality improvement and/or patient safety initiatives
	Our facility accesses targeted remote stewardship expertise
	Our stewardship team works with the microbiology laboratory to inform cascade and/or selective reporting protocols for isolate susceptibilities
	Our stewardship team monitors compliance with appropriate surgical prophylaxis
	Which committees or leadership entities provide oversight of your facility's antibiotic stewardship efforts?
	On average, what percent time does the physician (co) leader dedicate to antibiotic stewardship activities in your facility?
	On average, what percent time does the pharmacist (co) leader dedicate to antibiotic stewardship activities in your facility?
	What percent time for antibiotic stewardship activities is specified in the physician (co) leader's contract or job description?
	What percent time for antibiotic stewardship activities is specified in the pharmacist (co) leader's contract or job description?

Antimicrobial Stewardship Challenges

Feedback on the proposed rule specific to CAHs revealed concerns around limited staffing, infrastructure, and financial resources available for implementation of such requirements [2, 16]. According to the 2016 NHSN Annual Hospital Survey, 46% of CAHs had implemented all 7 CDC Core Elements [16]. Although increased from 26.3% the year prior, this was in stark contrast to the 64% implementation rate by large hospitals [16, 17]. Stewardship guidelines have recommended a stewardship team consist of an infectious diseases physician, infectious diseases-trained pharmacist, clinical microbiologist, information technology specialist, infection preventionist, and a hospital epidemiologist [18]. Inherently, CAHs have key differences compared to other community hospitals which can make ASP implementation a challenge. A study on ASP implementation in Kansas found statistically lower rates of CAHs reporting having a full-time pharmacist, support from leadership, clinical decision support, antibiograms, or a defined formulary when compared to the other acute care facilities [19]. Similar barriers identified in a literature review included the following:

- Lack of infectious diseases-trained clinicians
- Difficulty recruiting to small hospitals
- Geographical isolation
- Lack of information technology support
- Insufficient buy-in from leadership and staff
- Competing priorities

- Lack of pharmacy resources

Adapted from: Bishop et al. [20]

A final barrier of note is in measurement of antibiotic consumption. Without information technology support, collection of data may require manual review. Additionally, regardless of the metric utilized, antimicrobial utilization data can easily be skewed by low patient volumes at CAHs [21].

Solutions and Strategies

In 2017, the CDC published a guide for the Implementation of Antibiotic Stewardship Core Elements at Small and Critical Access Hospitals to supplement and target existing core elements resources to help address the lag in uptake [2]. Despite their unique challenges, some CAHs have successfully implemented core element compliant ASPs. As of 2015, >200 CAHs reported full implementation of the 7 CDC core elements. A review of published literature found that the following resources were mentioned as key assets for ASP implementation:

- Significant leadership support
- Identification of ASP champions
- Health-system or telemedicine resources
- Centralized microbiology, epidemiology, and infection prevention support
- Additional training for ASP team
- Collaborative, team-based approach to patient care
- Pride in facility

Adapted from: Bishop et al. [20]

These findings were consistent with tools discussed in the CDC Implementation guide [2]. A summary of those strategies as well as other recommendations from successful stewardship programs is presented in Table 4.3.

Table 4.3 Summary of strategies to support CAH implementation of each core element [3, 16, 21, 22]

Core Element	Strategies
Leadership	Dedicated time and resources remain important
	Align with existing quality improvement goals
	Utilize federal, state, or health-system financial resources as available
Accountability	Designate a senior executive or direct report as accountable for success
	Formal statement of support from hospital board
	Creation of ASP reporting structure
Pharmacy expertise	Utilize state collaborative or health-system resources as available
	Telemedicine
	Additional training for non-infectious diseases-trained physician and/or pharmacist
Action	Potential to review all patients on antibiotics due to low volume
	Focus on straightforward, guideline-based, or high-yield initiatives
	Target most common disease states
	Daily review of culture results
	Expanded role of support team, especially nurses
Tracking	Utilization of administration data
	Submit to NHSN antibiotic use option
	Utilization of defined daily doses (DDD) per 1000 patient-days as alternative consumption metric
	Incorporate regional antibiogram data
	Utilize state collaborative or health-system resources as available
Reporting	Regular reporting to hospital leadership
	Prescriber-specific report cards
	Standing agenda item on existing meetings
	All-staff electronic communication
Education	Incorporate into new hire and annual education for all employees
	Regular communication via intranet or newsletters
	One-on-one education as available
	Include information on patient discharge education

While the importance of leadership commitment remains the same, support for ASPs in CAHs may look different from a resource standpoint. Dedicated time and funding for ASP activities is still vital to success. A recent study found that stewardship activities at <70 bed hospitals take <5 h per week [21]. Leveraging partnerships that use part-time or contracted experts, accomplished through health-system or telemedicine resources can easily achieve these needs [16, 20]. Additionally, aligning ASP activities with existing administrative goals and collaborating with key stakeholders can also help build support and foster ongoing relationships [22, 23].

Lack of infectious diseases expertise is consistently cited as an unavailable on-site resource for CAHs. Healthcare systems, state collaboratives, consulting groups, and telemedicine have been utilized to overcome this barrier [11, 21, 23–26]. Models have included remote consultation via secure software, a weekly visiting specialist, or deferral to ASP coordination to an outside group within the health-system or even a group outside of the system [16, 18, 21, 27]. Successful ASPs have also been led by hospitalists and non-ID pharmacists [2, 21]. Robust stewardship training courses are available from multiple infectious diseases organizations to enhance skills of the ASP team as needed. Physician training in Antimicrobial Stewardship is available through the Infectious Diseases Society of America (IDSA) to ID fellows through the Core Antimicrobial Stewardship Curriculum. Training to non-specialists and ID physicians outside of fellowship is available through the CDC Training Course on Antimicrobial Stewardship [28]. Formal training for pharmacists is available through the Society of Infectious Diseases Pharmacists and Making a Difference in Infectious Diseases (MAD-ID) [29, 30].

The CDC implementation guide has specific recommendations for targets to consider for ASP initiatives. General guidance includes ensuring that the initiative and associated tracking is feasible, aligning with organizational priorities, and assessing baseline data to identify areas of opportunity. Targeting broad-spectrum antimicrobials, IV to PO conversion, and duration of antibiotic therapy are high-yield initial interventions for many ASPs. Targeting the most treated disease states, community-acquired pneumonia, urinary tract infection, and skin and soft tissue infection is also recommended [2, 25]. Incorporation of region-specific guidelines may help enhance these initiatives when antibiogram data is not readily available. Similarly, reviewing daily culture results is a valuable patient safety intervention. The implementation guide also highlights the expanded role of nurses in CAH ASPs particularly in relaying information from the frontline to the ASP team but also assisting with antibiotic time-outs [2]. Especially in remote settings, nurses remain at the bedside, contributing to triaging infectious symptoms and antibiotic adverse events, identifying opportunities for de-escalation or switch to oral medication, and educating patients [31].

One advantage of ASP initiatives at CAHs over larger institutions is the small number of providers and ability to personalize interventions. The concept of "Handshake Stewardship" is increasingly being recognized as a core element of successful ASPs. With fewer providers on site at the CAH, the likelihood of having face-to-face interactions with the ASP team is increased. The study of Handshake Stewardship was born in pediatric medicine and critical care medicine, which also capitalize on a smaller patient population with a smaller number of providers to interface with, drawing similarities to CAHs [32, 33]. With bidirectional information exchange, there are greater opportunities for education and improved quality [33].

Key differences also exist for tracking and reporting of stewardship outcomes. For CAHs without IT support, reporting to the NHSN AU module or changing from DOTs to the more easily calculated DDD metric is encouraged [2]. Quarterly reports rather than monthly may be advisable to reduce variability [21]. With few beds and low census, one or two additional patients consuming antibiotics may greatly skew the data set. However, more summative evaluations help wash away temporal anomalies. Reporting on cost-savings or disease-state specific guideline adherence has also been used effectively by CAH ASPs. The use of prescriber-specific report cards may be more feasible in the CAH setting due to staff size.

Finally, ongoing ASP education for hospital employees may be challenging at CAHs. Didactic or other on-site education may be limited by a lack of standing meetings or remote location. Similar to the other core elements, the CDC recommends incorporating education into existing resources such as onboarding education or patient discharge instructions [2]. Access to electronic resources for education can meet both clinical goals and regulatory requirements [14].

Conclusion

Regardless of the known barriers, implementation of fully complaint ASPs in CAHs is still vital to limiting suboptimal antimicrobial use in these facilities. It is advised to start with small projects that are likely to be successful. Careful assessment of existing administrative goals and facility resources will help identify feasible targets for the ASP. Even small changes can have a significant impact on patient outcomes, hospital-acquired infection, and C. *difficile* rates, in addition to satisfying regulatory standards. As more remote hospitals are successful in ASP implementation and leaders share their success stories, those ideas can be used to optimize antimicrobial use in resource-limited settings.

References

1. Rural Health Information Hub: Critical Access Hospitals; 2019. https://www.ruralhealthinfo.org. Accessed 25 Mar 2021.
2. Centers for Disease Control and Prevention. Implementation of antibiotic stewardship core elements at small and critical access hospitals. National Center for Emerging and Zoonotic Diseases; 2015. https://www.cdc.gov/antibiotic-use/healthcare/implementation/core-elements-small-critical.html. Accessed 25 Mar 2021.
3. National Rural Health Resource Center. MBQIP measures fact sheets. Stratis Health; 2019. https://www.ruralcenter.org/sites/default/files/MBQIP-Measures-Fact-Sheets-Jan-2019.pdf. Accessed 11 Feb 2019.
4. Medicare and Medicaid Programs; Regulatory Provisions to Promote Program Efficiency, Transparency, and Burden Reduction; Fire Safety Requirements for Certain Dialysis Facilities; Hospital and Critical Access Hospital (CAH) Changes to Promote Innovation, Flexibility, and Improvement in Patient Care, 84 (189) Fed. Reg. 51775-51776 (September 30, 2019). Critical Access Hospital Locations List. Flex Monitoring Team. www.flexmonitoring.org/critical-access-hospital-locations-list. Accessed 3 Mar 2021.
5. Fast Facts on US Hospitals, 2021. Chicago: American Hospital Association; 2021. www.aha.org/statistics/fast-facts-us-hospitals. Accessed 25 Mar 2021.
6. Stenehjem E, Hersh AL, Sheng X, Jones P, Buckel WR, Lloyd JF, et al. Antibiotic use in small community hospitals. Clin Infect Dis. 2016;63:1273–80. https://doi.org/10.1093/cid/ciw588.
7. Stenehjem E, Hersh AL, Pavia AT. Reply to Buetti et al. Clin Infect Dis. 2017;64:985–6. https://doi.org/10.1093/cid/cix041.
8. Couderc C, Lacavé L, L'hériteau F, Astagneau P. Surveillance of overall hospital antibiotic consumption: is stratification according to hospital size the best method? Infect Control Hosp Epidemiol. 2011;32:1223–5. https://doi.org/10.1086/662622.

9. MacDougall C, Polk RE. Variability in rates of use of antibacterials among 130 US hospitals and risk-adjustment models for interhospital comparison. Infect Control Hosp Epidemiol. 2008;29:203–11. https://doi.org/10.1086/528810.
10. Magill SS, Edwards JR, Beldavs ZG, Dumyati G, Janelle SJ, Kainer MA, et al. Prevalence of antimicrobial use in US acute care hospitals, May-September 2011. JAMA. 2014;312:1438–46. https://doi.org/10.1001/jama.2014.12923.
11. Howell CK, Jacob J, Mok S. Remote antimicrobial stewardship: a solution for meeting the joint commission stewardship standard? Hosp Pharm. 2019;54(1):51–6. https://doi.org/10.1177/0018578718769240.
12. Buetti N, Atkinson A, Kronenberg A, Marschall J. Different epidemiology of hospital-acquired bloodstream infections between small community hospitals and large community hospitals. Clin Infect Dis. 2017;64:984–5. https://doi.org/10.1093/cid/cix040.
13. Gandra S, Trett A, Klein EY, Laxminarayan R. Is antimicrobial resistance a bigger problem in tertiary care hospitals than in small community hospitals in the United States? Clin Infect Dis. 2017;65(5):860–3. https://doi.org/10.1093/cid/cix413.
14. CDC. Core elements of hospital antibiotic stewardship programs. Atlanta: US Department of Health and Human Services, CDC; 2019. Available at https://www.cdc.gov/antibiotic-use/core-elements/hospital.html. Accessed 25 Mar 2021, Jan 2019
15. National Healthcare Safety Network (NHSN). 57.103 Patient safety component—annual facility survey form January 2021. In: 5-Step enrollment for acute care hospitals/facilities. Centers for Disease Control and Prevention, National Center for Emerging and Zoonotic Infectious Diseases (NCEZID), Division of Healthcare Quality Promotion (DHQP); 2021. https://www.cdc.gov/nhsn/forms/57.103_pshospsurv_blank.pdf. Accessed 25 Mar 2021.
16. Bhatt J, Smith B, Neuhauser MM, Srinivasan A, Moore P, Hyun DY. Collaborative solutions to antibiotic stewardship in small community and critical access hospitals. Acad Med. 2019;94(10):1419–21. https://doi.org/10.1097/ACM.0000000000002859.
17. O'Leary EN, van Santen KL, Webb AK, Pollock DA, Edwards JR, Srinivasan A. Uptake of antibiotic stewardship programs in US acute care hospitals: findings from the 2015 National Healthcare Safety Network Annual Hospital Survey. Clin Infect Dis. 2017;65(10):1748–50. https://doi.org/10.1093/cid/cix651.
18. Dellit TH, Owens RC, McGowan JE Jr, Gerding DN, Weinstein RA, Burke JP, et al. Infectious Diseases Society of America and the Society for Healthcare Epidemiology of America guidelines for developing an institutional program to enhance antimicrobial stewardship. Clin Infect Dis. 2007;44:159–77. https://doi.org/10.1086/510393.
19. Poteete C, Scaletta JM. Antimicrobial stewardship in Kansas: results from a statewide survey. Am J Infect Control. 2017;45(1):42–5. https://doi.org/10.1016/j.ajic.2016.08.016.
20. Bishop J, Kong DC, Schulz TR, Thursky KA, Buising KL. Meeting the challenge for effective antimicrobial stewardship programs in regional, rural and remote hospitals - what can we learn from the published literature? Rural Remote Health. 2018;18:4442. https://doi.org/10.22605/RRH4442
21. Stenehjem E, Hyun DY, Septimus E, Yu KC, Meyer M, Raj D, Srinivasan A. Antibiotic stewardship in small hospitals: barriers and potential solutions. Clin Infect Dis. 2017;65:691–6. https://doi.org/10.1093/cid/cix407.
22. Buckel WR, Veillette JJ, Vento TJ, Stenehjem E. Antimicrobial stewardship in community hospitals. Med Clin North Am. 2018;102:913–28. https://doi.org/10.1016/j.mcna.2018.05.005.
23. Sexton DJ, Moehring RW. Implementation of antimicrobial stewardship programs in small community hospitals: recognizing the barriers and meeting the challenge. Clin Infect Dis. 2017;65(4):697–8. https://doi.org/10.1093/cid/cix409.
24. Stenehjem E, Hersh AL, Buckel WR, Jones P, Sheng X, Evans RS, et al. Impact of implementing antibiotic stewardship programs in 15 small hospitals: a cluster-randomized intervention. Clin Infect Dis. 2018;67:525–32. https://doi.org/10.1093/cid/ciy155.
25. Jenkins TC, Hulett T, Knepper BC, Shihadeh KC, Meyer MJ, Barber GR, et al. A Statewide antibiotic stewardship collaborative to improve the diagnosis and treatment of urinary tract

and skin and soft tissue infections. Clin Infect Dis. 2018;67:1550–8. https://doi.org/10.1093/cid/ciy268.
26. Yam P, Fales D, Jemison J, Gillum M, Bernstein M. Implementation of an antimicrobial stewardship program in a rural hospital. Am J Health Syst Pharm. 2012;69(13):1142–8. https://doi.org/10.2146/ajhp110512.
27. dosSantos RP, Deutschendorf C, Carvalho OF, Timm R, Sparenberg A. Antimicrobial stewardship in a community hospital in Southern Brazil. J Telemed Telecare. 2013;19:1–4. https://doi.org/10.1177/1357633X12473901.
28. CDC Training on Antibiotic Stewardship. Washington DC: Centers for Disease Control and Prevention; 2019. www.train.org/cdctrain/training_plan/3697. Accessed 15 Aug 2019.
29. Antimicrobial Stewardship Certificate programs. Geneva: Society of Infectious Diseases Pharmacists. www.sidp.org/Stewardship-Certificate. Accessed 15 Aug 2019.
30. Antimicrobial Stewardship programs. Fairfield: Making a Difference in Infectious Diseases. mad-id.org/antimicrobial-stewardship-programs/. Accessed 15 Aug 2019.
31. Olans RN, Olans RD, DeMaria A. The critical role of the staff nurse in antimicrobial stewardship- unrecognized, but already there. Clin Infect Dis. 2016;62(1):84–9. https://doi.org/10.1093/cid/civ697.
32. Hurst AL, Child J, Pearce K, Palmer C, Todd JK, Parker SK. Handshake stewardship. Pediatr Infect Dis J. 2016;35(10):1104. https://doi.org/10.1097/INF.0000000000001245.
33. Peermohamed S, Koscar J, Evans B. Embracing handshake stewardship: utility of collaborative, prospective audit and feedback rounds in an intensive care unit at Royal University Hospital, Saskatoon, Saskatchewan. Open Forum Infect Dis. 2017;4(S1):S272. https://doi.org/10.1093/ofid/ofx163.603.

Antimicrobial Stewardship in Ambulatory Surgery Centers

5

Sophia Macleay Cardwell, Andrew Root, and Andrea Halliday

Epidemiology of Surgical Site Infections

In the inpatient setting, surgical site infections (SSI) occur in 2–4% of patients undergoing surgery [1]. SSIs are the most common inpatient healthcare acquired condition tracked by the National Healthcare Safety Network (NHSN) [2]. In ambulatory surgery centers, data collection techniques for infection rates are in the process of being standardized through NHSN. Estimates using a variety of methods vary by procedure and have yielded rates ranging from lower than, comparable to, or higher than inpatient infection rates using NHSN criteria. Anywhere from 0% to 6% rates have been reported [3–5]. Infection control practices in ambulatory surgical centers also vary considerably, with survey data indicating lapses in infection control practices are common in some settings [6].

S. M. Cardwell (✉)
Department of Pharmacy, PeaceHealth Columbia Region, Vancouver, WA, USA
e-mail: scardwell@peacehealth.org

A. Root
Department of Infectious Diseases, The Vancouver Clinic, Vancouver, WA, USA
e-mail: aroot@tvc.org

A. Halliday
Department of Administration, Sacred Heart Medical Center RiverBend, Springfield, OR, USA

S. Doron, M. Campion (eds.), *Antimicrobial Stewardship in Non-Traditional Settings*, https://doi.org/10.1007/978-3-031-21710-4_5

Principles of Surgical Infection Prophylaxis

SSIs require both microbial contamination of a surgical wound and impaired host immune response. Surgical procedures are generally classified as clean, clean/contaminated, contaminated, and dirty, based on risk of bacterial contamination [7] (Table 5.1). Bacterial contamination will be present in all surgical wounds, regardless of technique and surgery type. Aseptic technique greatly minimizes, but does not eliminate, bacterial contamination of wounds. Greater inoculum of surgical wounds is associated with greater risk of SSI; making reduction of bacterial contamination a cornerstone of surgical site infection prevention. Modifiable risk factors for surgical site infections aimed at minimizing bacterial contamination include skin preparation, sterile technique, management of catheters or drains, and antimicrobial prophylaxis. Modifiable risk factors aimed at preservation of the host immune response include elements of peri-operative care such as glycemic control, perfusion, and temperature control. Non-modifiable risk factors for both bacterial contamination and host immune response include patient age, comorbidities, immune and functional status, as well as the type and duration of the surgical procedure.

Appropriate antimicrobial prophylaxis, in addition to other key elements of pre-operative care, is demonstrated to drastically reduce the incidence of SSI for a

Table 5.1 Surgical Wound Classification

Class I: Clean	An uninfected operative wound in which no inflammation is encountered and the respiratory, alimentary, genital, or uninfected urinary tract is not entered. In addition, clean wounds are primarily closed and, if necessary, drained with closed drainage. Operative incisional wounds that follow non-penetrating (blunt) trauma should be included in this category if they meet the criteria
Class II: Clean-contaminated	An operative wound in which the respiratory, alimentary, genital, or urinary tracts are entered under controlled conditions and without unusual contamination. Specifically, operations involving the biliary tract, appendix, vagina, and oropharynx are included in this category, provided no evidence of infection or major break in technique is encountered
Class III: Contaminated	Open, fresh, accidental wounds. In addition, operations with major breaks in sterile technique (e.g., open cardiac massage) or gross spillage from the gastrointestinal tract, and incisions in which acute, nonpurulent inflammation is encountered are included in this category
Class IV: Dirty-infected	Old traumatic wounds with retained devitalized tissue and those that involve existing clinical infection or perforated viscera. This definition suggests that the organisms causing post-operative infection were present in the operative field before the operation

Mangram, Alicia J.; Horan, Teresa C.; Pearson, Michele L.; Silver, Leah Christine; Jarvis, William R. (William Robert), 1948-; Guideline for the prevention of surgical site infection, 1999. Source: Infect Control Hosp Epidemiol. 1999 Apr;20(4):250–78; quiz 279–80 variety of procedures. Antimicrobial prophylaxis is not required for many low-risk clean procedures, but is recommended for certain high-risk clean procedures, as well as clean/contaminated, contaminated, and dirty procedures [7]. Agents used for antimicrobial prophylaxis should be active against likely pathogens of SSI for the procedure, dosed and administered in a manner ensuring adequate concentration at the surgical site from immediately prior to incision until wound closure, minimally toxic to the host and host microbiome, and given for the shortest effective period needed to minimize infection risk [7, 8].

Ambulatory surgery centers can improve effectiveness of antimicrobial prophylaxis by creating comprehensive tools to support appropriate antimicrobial prophylaxis. Facility and procedure-specific guidelines with process education around indication, selection, timing, and duration of antimicrobial prophylaxis are demonstrated to improve efficacy and minimize misuse of antimicrobials [9].

Timing of Antimicrobial Administration

Effective antimicrobial prophylaxis requires administration prior to incision. The previously recommended window of timing of antimicrobial administration was dependent on the infusion time of various antimicrobials used for prophylaxis, with those requiring 60 min or greater infusion time recommended to be administered within 120 min of incision, and others within 60 min of incision [8]. More recent recommendations include simply the adage that sufficient concentration at the surgical site be achieved by the time of incision and throughout the procedure.

The time of administration of antimicrobial prophylaxis has demonstrable importance to avoidance of SSI (Fig. 5.1). A variety of studies have examined the relationship of timing of prophylaxis to rates of SSI, noting that the risk of SSI among patients receiving cephalosporins for antimicrobial prophylaxis is highest for patients whose antimicrobials were administered after the incision is made and for those receiving prophylaxis greater than 120 min prior to incision. Rates are lowest when prophylaxis is initiated within the 60 min prior to incision [10–12] (Fig. 5.1). For vancomycin, rates of SSI are lowest within a similar time frame, but notably higher if within the 15 min prior to incision [13] (Fig. 5.1). Other research examining vancomycin used for pre-operative prophylaxis attributes unacceptably high rates of SSI with administration too near to incision time, thus not allowing adequate concentration at the surgical site prior to wound contamination [14].

In order to maximize the concentration of antimicrobial at the site of incision, timing prophylactic antimicrobials to finish infusing immediately prior to the time of incision is reasonable.

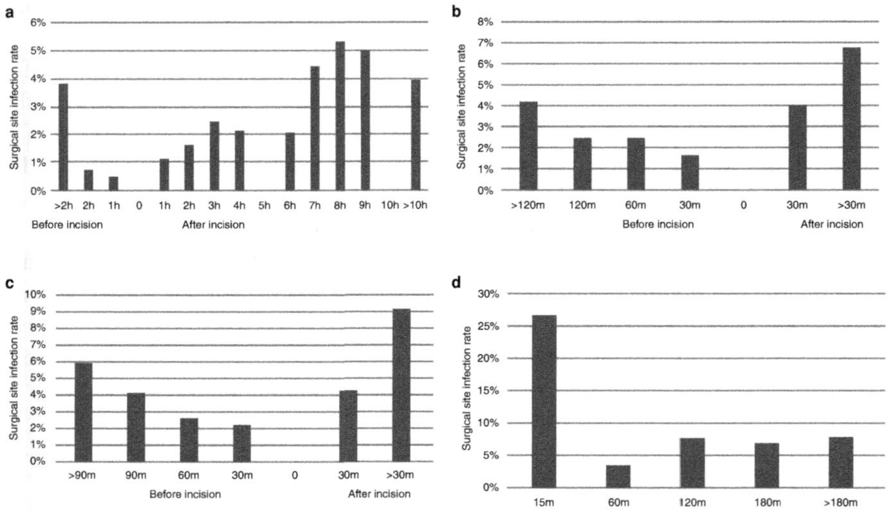

Fig. 5.1 Timing of antimicrobial prophylaxis and infection rate. Lowest infection rates demonstrated with end of infusion immediately prior to incision, regardless of drug class. Cephalosporin (3–30 min length infusion) prophylaxis peri-operatively (**a**–**c**) and vancomycin (60+ min infusion) (**d**) to SSI rate. Adapted from Classen [10] (**a**), Steinberg [11] (**b**), van Kasteran [12] (**c**), and Garey [13] (**d**)

Antimicrobial Agents

Effective antimicrobials for surgical prophylaxis must have reliable activity against common pathogens, without unnecessarily broad activity against unlikely pathogens of SSI. Common pathogens for all surgical procedures include *Staphylococcus aureus* (MSSA and MRSA), coagulase negative *Staphylococci* (e.g., *Staphylococcus epidermidis*), and beta hemolytic *Streptococcus* spp. (e.g., *Streptococcus pyogenes* and *S. agalactiae*). Additional activity against gram negative or anaerobic pathogens is necessary for some gastrointestinal and genitourinary procedures. Activity against these pathogens by agents approved for surgical prophylaxis will vary considerably by geography, patient population, and local antimicrobial use patterns. It is important to incorporate local susceptibility patterns into formulary decisions regarding recommended agents for pre-operative prophylaxis. Antibiograms developed by area hospitals, infectious diseases or public health organizations, or research institutions are ideal resources for local susceptibility data.

Successful prophylaxis is dependent on the activity of the administered agent against likely pathogens. For example, even though clindamycin is a recommended agent for pre-operative prophylaxis, susceptibility rates for *S. aureus* are as low as 50% in some parts of the country. Utilizing tools such as an antibiogram with local susceptibility data is vital to help determine which of the available agents is best matched against likely pathogens in a population and should be used when making prophylaxis recommendations.

Among agents available for surgical prophylaxis, considerable variability exists between spectrum of activity, resistance profile, timing and administration issues, and toxicity (Table 5.2) [15]. Selecting a formulary for use in ambulatory surgery centers should account for these differences in order to maximize efficacy and safety and minimize excessive toxicity, cost, and case time.

Table 5.2 Properties of select antimicrobial agents approved for use as surgical prophylaxis

Drug	Dose (IV)	Half-life normal renal function (h)	Re-dosing interval (h)
Ampicillin/sulbactam	3 g	0.8–1.3	2
Piperacillin/tazobactam	3.375 g	0.7–1.2	2
Cefazolin	2–3 g	1.2–2.2	4
Cefuroxime	1.5 g	1–2	4
Cefoxitin	2 g	0.7–1.1	2
Cefotetan	2 g	2.8–4.6	6
Vancomycin	15 mg/kg dosing weight (1–2 g)	6–8	Not usually indicated
Clindamycin	900 mg	2–4	6
Ciprofloxacin	400 mg	3–7	Not usually indicated
Levofloxacin	500 mg	6–8	Not usually indicated
Gentamicin	5 mg/kg dosing weight	2–3	Not usually indicated[a]
Metronidazole	500 mg	6–10	Not usually indicated

[a]A half-life of 2 h would indicate consideration be given for re-dosing if a procedure exceeded 4 h, however the post-antibiotic effect of aminoglycosides supersedes this recommendation

Dosing of Agent

Data regarding efficacy of weight-based dosing of prophylactic antimicrobials are sparse. Dosing recommendations are based on measured tissue concentrations in experimental settings with single administration. Because of this, recommended surgical prophylactic doses are often higher than treatment doses, by virtue of needing to attain maximal tissue concentration after a single dose. Pharmacokinetic analysis of the need for weight-based dosing in obese patients is also inconclusive [16, 17]. The United States Centers for Disease Control and Prevention (CDC) guidance makes no recommendation regarding weight-based dosing for obese patients, citing a lack of evidence [7]. A variety of dosing recommendations are available; weight-based dosing for obese patients is increasingly common [8].

Unlike weight-based dosing, the timing of dose and spectrum of activity are definitively linked to prevention of SSI. Thus, a reasonable approach to dosing by weight is to standardize recommendations for an individual facility or group of facilities based on multidisciplinary input, in order to ensure confusion about dosing schemes does not interfere with optimal selection and timing of antimicrobial prophylaxis.

Similar to the need for weight-based dosing, intra-operative re-dosing for prolonged surgeries (greater than 4 h) has limited evidence. It is recommended by surgical, pharmacy, and infectious diseases societies, but does not have a solid evidence base and is not specifically recommended by the CDC. Nevertheless, re-dosing recommendations based on an estimation of two times the half-life of the drug for prolonged duration of surgery are readily available (Table 5.3) and should also be standardized in an institution to avoid confusion regarding antimicrobial administration.

Table 5.3 Dose and re-dosing interval of select antimicrobial agents approved for use as surgical prophylaxis (Modified from Bratzler [8])

Drug	Spectrum of activity (absent acquired resistance)	No activity	Administration time	Major toxicity	Resistance/use profile
Cephalosporins: Cefazolin Cefuroxime	MSSA, coagulase negative *Staphylococcus* spp., *Streptococcus* spp., enterobacterales	MRSA, *enterococcus* spp., colonic anaerobes (*Bacteroides* spp.)	30 min via infusion, 3–5 min via slow IV push	Minimal	Resistance approaches 40% for coagulase negative *Staphylococcus* spp., 30% for enterobacterales, refer to local data
Cephamycins: Cefoxitin Cefotetan	MSSA, coagulase negative *Staphylococcus* spp., *Streptococcus* spp. enterobacterales, colonic anaerobes (*Bacteroides* spp.)	MRSA, *Enterococcus* spp.	30 min via infusion, 3–5 min via slow IV push	Minimal	Resistance approaches 40% for coagulase negative *Staphylococcus* spp., 20% for colonic anaerobes, refer to local data
Vancomycin	MSSA, MRSA, coagulase negative *Staphylococcus* spp., *Streptococcus* spp.	Gram negative organisms	60–120 min, dose dependent; IV push not recommended due to toxicity	Associated with greater rates of AKI than cephalosporins	Sub-optimal activity against MSSA compared to cephalosporins
Clindamycin	MSSA, MRSA, coagulase negative *Staphylococcus* spp., *Streptococcus* spp. colonic anaerobes (*Bacteroides* spp.)	Gram negative organisms, *Enterococcus* spp.	30 min via infusion; IV push not recommended due to toxicity	Associated with greater rates of *C. difficile* than other agents	Resistance approaches 50% for all pathogens dependent on population, refer to local data; sub-optimal activity against gram positive organisms compared to cephalosporins and vancomycin

(continued)

Table 5.3 (continued)

Drug	Spectrum of activity (absent acquired resistance)	No activity	Administration time	Major toxicity	Resistance/use profile
Fluoroquinolones: Ciprofloxacin Levofloxacin	MSSA, MRSA coagulase negative *Staphylococcus* spp., *Streptococcus* spp., enterobacterales	Colonic anaerobes (*Bacteroides* spp.)	60 min via infusion; IV push not recommended due to toxicity	Associated with greater rates of *C. difficile*, impaired healing of fractures	Resistance among *Staphylococcus* spp. widespread, approaches 40% for enterobacterales, refer to local data; sub-optimal activity against gram positive organisms compared to cephalosporins and vancomycin
Gentamicin	Enterobacterales	Gram positive organisms, *Enterococcus* spp., colonic anaerobes (*Bacteroides* spp.)	30 min via infusion; IV push not recommended due to toxicity	Nephrotoxic	Sub-optimal against preventing SSI compared to beta-lactam prophylaxis
Metronidazole	Colonic anaerobes (*Bacteroides* spp.)	Aerobic organisms	30 min via infusion; formulation precludes IV push	Minimal	Require combination with agent active against aerobic pathogens

MSSA Methicillin sensitive *S. aureus*, MRSA Methicillin resistant *S. aureus*

Penicillin Allergy

Beta-lactam antimicrobials (penicillins, cephalosporins, and carbapenems) are some of the safest and most effective for treatment and prevention of infection. First and second generation cephalosporins, primarily cefazolin, are a mainstay of therapy for surgical infection prophylaxis due to their ideal spectrum of activity, safety, and toxicity profile. In some institutions, cephalosporins are avoided in patients with reported penicillin allergy, out of concern of cross reactivity. An often-cited rate of 10% cross reactivity between penicillin allergy and cephalosporin allergy appears to stem from research in the 1970s, where these rates were noted in patients with self-reported penicillin allergy administered cephalosporins [18, 19]. These studies had some important structural flaws, most notably that the administered cephalosporins contained penicillin [20].

More recent understanding of cross reactivity concludes that allergic cross reactivity is related to similarities between molecular side chains, rather than the core beta-lactam structure [20, 21]. Rates of cross reactivity between molecules with similar side chains, such as ampicillin and cephalexin, are indeed elevated. Rates of cross reactivity of penicillin allergy with unrelated cephalosporins such as cefazolin, cefuroxime, ceftriaxone, and cefepime are minimal, and similar to rates in the general, non-penicillin allergic population [20–22]. Additionally, it is worth noting that patients with a true penicillin allergy are at an increased risk of allergy to any other, unrelated drug, by as much as threefold [23]. As such, those infrequent allergies to unrelated cephalosporins in patients with penicillin allergy may be considered to be co-reactive, rather than cross-reactive [21].

Approximately 10% of the US population reports a penicillin allergy [24]. This number does not accurately represent the rate of true allergy, for several reasons. Self-reported "allergies" are often consistent with adverse drug effects such as gastrointestinal complaints, viral exanthem, headache, and the like. As much as 90% of self-reported penicillin allergies are not true allergic reactions. The incidence of true IgE mediated penicillin hypersensitivity is 0.02–0.04%. Additionally, true penicillin hypersensitivity wanes over time, at a rate of 10% per year. Thus, over a 10-year period, 80–100% of patients with true hypersensitivity will no longer be penicillin allergic. A large cohort study examining penicillin allergy rates by skin test of only those patients with a self-reported penicillin allergy noted actual skin test positive rates of less than 2% [25]. Nevertheless, the practice of avoiding administration of beta-lactam antimicrobials for this population affects a significant number of patients. A variety of studies have demonstrated the detrimental effects of administration of non-beta-lactam antimicrobials for patients with penicillin allergy documented in their electronic medical record.

A large matched cohort study examining acquisition of penicillin allergy into a medical record noted a significant increase in utilization of non-beta-lactam antimicrobials, including macrolides, clindamycin, and fluoroquinolones. Over a 10-year period, a 69% increase in MRSA acquisition and a 26% increase *Clostridioides difficile* infection were documented, as compared to patients without a penicillin allergy [26].

Another matched cohort study noted significantly more beta-lactam alternative antimicrobials vancomycin, clindamycin, and fluoroquinolones, in patients with a documented penicillin allergy over a 2-year period. Significant increases were noted in MRSA and *C. difficile* rates (14.1 and 23.4%, respectively), as well as 30.1% more vancomycin-resistant *Enterococcus* spp. Those with penicillin allergy documented spent an average of 0.59 additional inpatient hospital days during the approximately 2-year follow-up period, associated with substantial cost [25].

Documented penicillin allergy and use of beta-lactam alternatives is also linked to increases in SSI. A retrospective cohort study examining the effect of a reported penicillin allergy noted beta-lactam alternative antimicrobials clindamycin, gentamicin, and vancomycin were administered for surgical prophylaxis much more frequently in patients with a documented penicillin allergy as compared to those with none. A 50% increase in odds of SSI was also associated with penicillin allergy. The increased odds of SSI was entirely attributable to the administration of beta-lactam alternative agents in this population [14]. Another retrospective cohort evaluating SSI in patients with a self-reported beta-lactam allergy noted similarly elevated rates of clindamycin, gentamicin, and vancomycin as compared to patients without reported allergies, who received predominantly cefazolin. SSIs were significantly more likely in the self-reported beta-lactam allergy group (7.3% vs. 4.8%), a risk which was again entirely attributable to receipt of alternative antibiotics on regression [27].

Addressing the issue of reported penicillin allergy in ambulatory surgery patients allows a greater portion of patients to routinely receive beta-lactam antimicrobials for pre-operative prophylaxis. This provides the opportunity to lower rates of SSI, as well as minimizing antimicrobial related complications such as acquisition of *C. difficile* or drug resistance among normal flora. There are several approaches for antimicrobial stewardship relating to penicillin allergy.

De-labeling patients as penicillin allergic based on a brief interview with a healthcare provider as part of pre-operative evaluation is a simple, low-cost, and high-yield method of minimizing barriers to administration of beta-lactam antimicrobials as pre-operative prophylaxis (Fig. 5.2). A pharmacist led de-labeling protocol is demonstrated to safely de-label 80% of patients based solely on an interview, with an additional 16% de-labeled after an oral penicillin challenge [29].

Protocols for oral challenge and skin testing are also cost effective and reliable. Removal of allergy via testing or direct challenge in pre-operative appointments is demonstrated to improve rates of beta-lactam use for prophylaxis and reduce costs and intra-operative time [30, 31].

Lastly, simply administering cephalosporins chemically dissimilar to penicillin such as cefazolin or cefuroxime to penicillin allergic patients is a strategy broadly supported by both experimental and clinical data [21, 22, 32]. Removing alerts of cross reactivity between penicillin allergic patients and unrelated cephalosporin administration in electronic medical record systems is a reasonable way to facilitate this process [33].

Addressing documented penicillin allergy is ideal opportunity for antimicrobial stewardship in the setting of ambulatory surgery centers for several reasons. The

Ensure patients completing this screening have NONE of the following contraindications

☐Taking a beta blocker
☐Taking an antihistamine
☐Pregnant
☐Any other medical condition that would not tolerate anaphylaxis

Name ______________________ DOB ____________ MRN______________

Procedure date____________ Surgeon ____________________

Penicillin allergy short questionnaire	**yes**	**no**
1. When did your penicillin reaction occur? Check yes if within the last year, all other answers check no		
2. If you had a rash, did it last less than 24 hours? If no rash or if unknown duration, check no.		
3. Did you require medical attention such as epinephrine, ED visit, ICU visit, or intubation?		
4. Did your reaction involve blistering, ulcers, sloughing of your skin or the lining of your mouth, eyes, or genitals, and/or were you diagnosed with Stevens Johnson syndrome (SJS) or toxic epidermal necrolysis (TENs)?		
5. Did your reaction involve any organ dysfunction or failure, or were you diagnosed with serum sickness, drug rash with eosinophilia and systemic symptoms (DRESS), or acute interstitial nephritis (AIN)?		
6. Have you tolerated any penicillin since your reaction? Penicillin (Pen VK), Amoxicillin (Amoxil), amoxicillin/clavulanate (Augmentin), ampicillin (Omnipen, Principen), ampicillin/sulbactam (Unasyn), piperacillin/tazobactam (Zosyn)		

If patient answers yes to 6 (has tolerated any penicillin since their reported reaction) they are CLEARED. No testing indicated, remove allergy from EMR.

If all questions 1-5 are NO, patient is LOW RISK. Provide informed consent form and proceed to oral challenge. Observe for 1 hour. Remove allergy from EMR if no reaction. Treat any systemic reaction per protocol.

If any answers to 1-3 are YES, patient is at HIGHER RISK. Recommend skin testing, provide allergy referral.

If answers 4 or 5 are YES, continue to avoid penicillins.

Fig. 5.2 Sample process for patient interviews facilitating penicillin allergy de-labeling or oral challenge and subsequent allergy de-labeling. Adapted from Staicu et al. [28]

sizeable morbidity and cost associated with administration of alternative (non-beta-lactam) antimicrobials, combined with marked over reporting of penicillin allergy in electronic medical records makes this a high impact initiative. Confusion among healthcare providers regarding the nature of cross reactivity between penicillin allergy and cephalosporin administration allows for interventions to be relatively simple; with clinical education targeted at increasing efficiency and reducing waste and adverse effects of antimicrobials.

Duration of Antimicrobial Prophylaxis

The surgical care improvement project (SCIP) of 2006 included three performance measures related to peri-operative antimicrobial administration; specifically, the time of initiation, spectrum of activity, and time of discontinuation. The latter performance measure noted that prophylactic antimicrobials should be discontinued *within* 24 h but was broadly interpreted as a recommendation to continue prophylaxis *for* 24 h.

Data concerning the efficacy of peri-operative prophylaxis as related to SSI demonstrate that timing, dose, and spectrum of antimicrobial can affect rates of SSI. Evidence supporting antimicrobial administration after wound closure for clean, clean-contaminated, and contaminated procedures is lacking. A large portion of trials examining the efficacy of pre-operative antimicrobial prophylaxis utilized a single pre-operative dose. Those including post-operative doses reported similar rates of SSI as single-dose prophylaxis, with increases in organ toxicity, adverse drug events, *C. difficile* infection, and acquisition of antimicrobial resistance among normal flora with prolonged use of prophylactic antimicrobials. Thus, the recommendation to discontinue prophylaxis within 24 h was aimed at reducing adverse effects of prolonged antimicrobials, rather than any additional protection against SSI with post-operative doses [8].

Since the initiation of SCIP inpatient measures, several organizations have released clinical practice guidance related to peri-operative antimicrobial prophylaxis (Table 5.4) [7, 8, 34–36]. All support single-dose prophylaxis, none recommend administration of antimicrobials after wound closure. The CDC guideline specifically recommends against administration of prophylactic antimicrobials after wound closure for clean or clean-contaminated procedures, even in the presence of a drain [7].

Some data are available regarding acquired resistance with prophylactic antimicrobial use in the surgical setting. A prospective cohort study examining the relationship between prolonged prophylaxis, SSI, and acquired resistance found no increase in SSI with prolonged prophylaxis, but a significant increase in acquired resistance in a 3-month follow-up [37]. A more recent study retrospectively examining patients with SSI found no link to pre- and intra-operative prophylaxis only (not prolonged) and acquired resistance within a 30-day follow-up [38]. It is reasonable to limit surgical prophylaxis to pre- and intra-operative doses as a means of reducing acquisition of antimicrobial resistance among normal flora.

C. difficile infection is associated with any antimicrobial use, including single-dose surgical prophylaxis, however higher rates are associated with prolonged prophylaxis. A prospective cohort of cardiovascular surgery patients identified prolonged prophylaxis as an independent risk factor for *C. difficile* infection in a 2-month follow-up [39]. Another large retrospective cohort identified prolonged prophylaxis as an independent risk factor for *C. difficile*, with each additional day of antimicrobial administration contributing significantly to increased risk [40]. Prolonged prophylaxis was not associated with a reduction in SSI in either study.

Table 5.4 Summary of recent recommendations regarding single versus multiple dose surgical antimicrobial prophylaxis

Organization	Date	Statement on single-dose prophylaxis	Statement on multiple-dose prophylaxis
CDC[a]	2017	Do not administer additional prophylactic antimicrobial agent doses after the surgical incision is closed in the operating room, even in the presence of a drain	No statement
ACS/SIS[b]	2016	Antibiotics should be discontinued at the time of incision closure	There is no evidence that prophylactic antibiotic administration after incision closure further decreases SSI risk across a range of procedures. Ongoing antibiotic administration increases the risk of *C. difficile* infection
WHO[c]	2016	Single-dose prophylaxis is non-inferior to prolonged for preventing SSIs	The panel recommends against the prolongation of surgical antibiotic prophylaxis after completion of the operation for the purposes of preventing SSI. Serious concerns for the risk of antimicrobial resistance exist for prolonged prophylaxis
SHEA/IDSA/AHA/APIC[d]	2014	Prophylactic antibiotics should be discontinued at the time of wound closure	No evidence antibiotics given after closure contribute to efficacy. Antibiotics administered after wound closure contribute to *C. difficile* disease and antibiotic resistance
IDSA/SIS/SHEA/ASHP[e]	2013	Limiting the duration of antimicrobial prophylaxis to a single dose can reduce the risk of *C. difficile* disease	Evidence is mounting that post-operative antimicrobial administration is not necessary for most procedures. Prolonged prophylaxis is associated with an increased risk of acquired antimicrobial resistance

[a]United States Centers for Disease Control and Prevention
[b]American College of Surgeons and Surgical Infection Society
[c]World Health Organization
[d]Society for Healthcare Epidemiology of America, Infectious Diseases Society of America, American Hospital Association, and the Association for Professionals in Infection Control and Epidemiology
[e]Infectious Diseases Society of America, Surgical Infection Society, Society for Healthcare Epidemiology of America, American Society of Health-System Pharmacists

Lastly, acute kidney injury (AKI) has recently been shown to be strongly associated with prolonged surgical prophylaxis. In a large cohort of surgical patients, antimicrobial prophylaxis beyond 24 h was associated with additional risk of AKI for every additional day of antimicrobial prophylaxis. This relationship was present for both cardiac and non-cardiac procedures, with no corresponding reduction in SSI for either group with prolonged prophylaxis [40].

Efficacy of SSI prevention is similar between single-dose prophylaxis and continuation of prophylaxis after wound closure, for all studied clean and clean/contaminated procedures. As rates of ambulatory surgical procedures increase for procedures such as total joint arthroplasty, emerging data continue to support the efficacy of single-dose prophylaxis [41, 42] Antimicrobials administered after wound closure are demonstrated to increase rates of organ toxicity, *C. difficile*, and acquired antimicrobial resistance. Ensuring antimicrobials, parenteral or oral, are not continued after wound closure for clean and clean-contaminated procedures is a simple, cost-effective, and evidence-based antimicrobial stewardship initiative for the ambulatory setting.

The Role of Microbiological Screening and Decolonization

Microbiological screening and decolonization prior to surgery is a logical approach to minimize bacterial burden prior to surgery, but data supporting use are mixed with regard to effectiveness by surgery type, pathogen, screening site, and methodology.

The benefit of pre-operative nasal screening for MRSA and MSSA and subsequent decolonization prior to incision in cardiothoracic procedures involving a sternal wound is well established [8, 43]. These data were promising enough to prompt research into similar application in joint arthroplasty. Screening and decolonization for MRSA and MSSA are also broadly supported for joint arthroplasty, in both inpatient and outpatient settings [8]. A prospective cohort examining rates of SSI after screening for MSSA and MRSA in elective joint arthroplasty demonstrated a 69% decrease in rates of SSI after positive patients were decolonized with a combination of nasal mupirocin and chlorhexidine skin wipes [44]. Decolonization protocols for *S. aureus* positive patients vary by institution; no one procedure is recommended above others. The World Health Organization simply recommends peri-operative intranasal application of mupirocin 2%, with or without a combination of chlorhexidine body wash, for cardiothoracic and orthopedic procedures in carriers of *S. aureus* [36]. The role of screening and decolonization for other procedures is less clear.

Pre-operative urine screening and subsequent treatment of asymptomatic bacteriuria (ASB) is common in some practice settings. For genitourinary procedures involving mucosal trauma, screening for and treating ASB prior to surgery is recommended [45]. These procedures carry a substantial risk for post-operative sepsis when performed in the presence of bacteriuria or candiduria. A pre-operative urine culture should be obtained in order to provide definitive, rather than empirical, treatment prior to surgery. A single dose of an active agent, administered immediately prior to the procedure, is reasonable.

For non-genitourinary procedures, screening for and treatment of ASB as surgical prophylaxis should be avoided. The practice, however, is commonplace in some institutions, particularly for joint arthroplasty. A variety of studies addressing this practice have confirmed that while a positive pre-operative urine culture may be a marker for increased risk of SSI in total joint arthroplasty patients, providing antimicrobials does not mitigate the risk [46–49]. Additionally, treating asymptomatic pre-operative urine cultures is associated with increased risk of adverse drug events, *C. difficile*, and acquired resistance.

To assess SSI rates related to treatment of ASB, a prospective cohort of patients was screened for ASB prior to total hip and total knee arthroplasty, then followed for 1 year. Patients with ASB were more likely to have an SSI in the first post-operative year (4.3% vs. 1.4%, $p = 0.001$), however there was no statistical significant difference in SSI rate between those receiving treatment for ASB and those with no treatment (3.9% vs. 4.7%, $p = 0.78$). Additionally, while SSI pathogens in patients with ASB were more likely to be gram negative than in those without, the isolates were not correlated with the initial isolates from the ASB screening [46].

Another large cohort examining rates of SSI based on pre-operative ASB found no association between pre-operative ASB and SSI. Similar to previous research, receipt of an antimicrobial with activity against the urinary isolate had no effect on SSI rates [50].

Screening for pyuria is also common in non-urologic surgical patients, with some attempting to utilize pyuria as an indicator of the need for antimicrobials as part of pre-operative prophylaxis. Generally, the presence, absence, or degree of pyuria is not helpful in determining the need for antimicrobials [45]. A retrospective cohort of surgical patients assessed the rate of pyuria without bacteriuria identified in pre-operative screening, whether the presence of pyuria was correlated with antimicrobial use, and whether screening and treating had any effect on SSI, UTI, or *C. difficile* rates post-operatively. Patients with pyuria were more likely to receive antimicrobials, and the odds of treatment also increased with increasing degree of pyuria. Pyuria was not associated with SSI, regardless of treatment, but was correlated with post-operative *C. difficile* infection. *C. difficile* rates were higher in patients who had received antimicrobials for prophylaxis based on pyuria, as well. Pyuria was also associated with increases in post-operative UTI in total joint arthroplasty patients, the risk of which was not mitigated by pre-operative antimicrobial treatment for pyuria [51].

Based on the preponderance of evidence demonstrating the lack of benefit and associated adverse effects and cost of pre-operative screening for ASB and pyuria, a cohort of total joint arthroplasty patients was examined after the removal of pre-operative urine screening. SSI rates were unchanged, coupled with substantial decreases in urinalysis with culture and associated peri-operative courses of antimicrobials. None of the pathogens implicated in SSIs in the post implementation period were urinary pathogens [52]. The institution noted a sizeable cost savings, without any corresponding increase in SSI or other complications.

Screening and treating for ASB and pyuria remain common in ambulatory surgery, particularly for orthopedic procedures. This practice is a not evidence based, adding cost and unnecessary exposure to antimicrobials without corresponding reductions in SSI. As such, eliminating this practice from ambulatory surgical procedures is an optimal antimicrobial stewardship initiative.

The Role of Antiseptic or Antibiotic Irrigations for Surgical Wounds

Antiseptic and antibiotic irrigations of clean or clean-contaminated surgical wounds are routinely used in ambulatory surgical centers, but evidence supporting their use is lacking. One such preparation, bacitracin irrigation, was recently removed from the market by the FDA for its lack of demonstrated efficacy and its known toxicity profile [53]. Available trials consist of assessment of various combinations of antimicrobial preparations, usually for a small number of patients and a specific procedure. These data are often not easily extrapolated to other procedures, and numbers are small enough that efficacy is not reliably demonstrated. Available clinical practice guidelines either do not recommend or make no recommendation on the use of antimicrobial irrigations, citing a lack of data [7, 35, 36].

As with any area of clinical practice where antimicrobials are frequently used without ample supporting evidence, opportunities for stewardship exist. The theoretical benefit of antimicrobial irrigation solutions is reducing SSI risk, particularly in bowel and genitourinary procedures, and those with implantable devices. Theoretical risks are local toxicity or hypersensitivity, common with ingredients such as bacitracin and neomycin, and development of resistance among normal flora. Hypersensitivity reactions to antimicrobial irrigation solutions are relatively rare but well documented [54, 55]. Lastly, development of resistance to antimicrobials in irrigation solutions or even harboring multidrug resistant pathogens within solutions used for irrigation are documented risks [56, 57].

Given the lack of data for efficacy and potential risks and additional cost associated with use, a logical antimicrobial stewardship initiative would be to assess the evidence base for current practice in an ambulatory surgery center. Potential exists to minimize the number of agents used, minimize the type of procedures in which antimicrobial irrigations are used, or simplify antimicrobial recipes to a standard antiseptic such as chlorhexidine or povidone iodine [56].

Clinician Education

Antimicrobial stewardship in any setting often requires clinical education regarding potential changes in practice. Antimicrobial use varies widely by practice site, geographic location, and fundamental knowledge of principles of anti-infective agents. Many clinicians are reluctant to change practices that have remained static for years. When providing clinical education, the importance of noting that changing current practice does not imply previous practices were negligent or harmful cannot be overstated. In modern medicine, clinicians are scrutinized in all aspects of care, with the aim being increasingly more efficient, faster, better, and safer at all times. Along the way, best or standard practices have been reassessed and foundations have changed. The more knowledge gained, the more important it becomes to embrace the opportunity to leave behind what was once good for what is now known to be better. Focusing on the shared goal of improving care is paramount, in our ongoing quest for excellence.

Principles of Antimicrobial Stewardship

Antimicrobial stewardship in ambulatory surgery centers should focus on analysis of current evidence for use of antimicrobials. The key principle to implementing practice change is understanding the risks and benefits associated with surgical prophylaxis, as well as which variables surrounding the administration of antimicrobials have the ability to change outcomes and which do not.

When administered optimally, antimicrobials drastically reduce the risk of infection. However, misused or delayed antimicrobials have almost no effect on rates of SSI, and prolonged, unneeded, or overly broad in spectrum agents cause significant harm, even when used with the best of intention. Antimicrobials are one part of a host of interventions to be made to increase the safety of surgery, but are a poor substitute for other quality measures. They do not modify risk associated with procedures after the initial pre-operative dose and are similarly unable to mitigate negative outcomes associated with chronic illness or poor living conditions. Attempting to use antimicrobials for this purpose should be avoided.

Focusing stewardship initiatives toward ensuring the maximum number of patients possible are able to receive first-line prophylaxis and that prophylaxis is both initiated and discontinued appropriately is an ideal first step, prioritizing safety. Eliminating unnecessary steps such as screening and treating urine for unrelated procedures or the continuation of prophylactic antimicrobials post-operatively builds on these principles of stewardship, minimizing harm. Standardization of practice in this manner has great potential to reduce risk of SSI and associated adverse effects of antimicrobials, while streamlining care and minimizing cost.

References

1. Hall MJ, Shwartzman A, Zhang J, Liu X. Ambulatory surgery data from hospitals and ambulatory surgery centers: United States 2010. Natl Health Stat Rep. 2017;102:1–15.
2. Cullen KA, Hall MJ, Golosinskiy. Ambulatory surgery in the United States. Natl Health Stat Rep. 2009;11:1–25.
3. Barie PS. Infection control practices in ambulatory surgical centers. JAMA. 2010;303(22):2295–7.
4. Rhee C, Huang SS, Berrios-Torres SI, Kaganov R, Bruce C, Lankiewicz J, et al. Surgical site infection surveillance following ambulatory surgery. Infect Control Hosp Epidemiol. 2015;36(2):225–8.
5. Vilar-Compte D, Roldan R, Sandoval S, Corominas R, de la Rosa M, Gordillo P, et al. Surgical site infections in ambulatory surgery: a 5 year experience. Am J Infect Control. 2001;29(2):99–103.
6. Schaefer MK, Jhung M, Dahl M, Schillie S, Simpson C, Llata E, et al. Infection control assessment of ambulatory surgical centers. JAMA. 2010;303(22):2273–9.
7. Berrios-Torres SI, Umscheid CA, Bratzler DW, Leas B, Stone EC, Kelz RR, et al. Centers for Disease Control and Prevention guideline for the prevention of surgical site infection, 2017. JAMA Surg. 2017;152(8):784–91.
8. Bratzler DW, Dellinger EP, Olsen KM, Perl TM, Auwaerter PG, Bolon MK, et al. Clinical practice guidelines for antimicrobial prophylaxis in surgery. Am J Health Syst Pharm. 2013;70:195–283.

9. Alerany C, Company D, Monterde J, Semeraro C. Impact of local guidelines and an integrated dispensing system on antibiotic prophylaxis quality in a surgical centre. J Hosp Infect. 2005;60:111–7.
10. Classen DC, Evans RS, Pestotnik SL, Horn SD, Menlove RL, Burke JP. The timing of prophylactic administration of antibiotics and the risk of surgical-wound infection. NEJM. 1992;326(5):281–6.
11. Steinberg JP, Braun BI, Hellinger WC, Kusek L, Bozikis MR, Bush AJ, et al. Timing og antimicrobial prophylaxis and the risk of surgical site infections: results from the trial to reduce antimicrobial prophylaxis errors. Ann Surg. 2009;250(1):10–6.
12. van Kasteran MEE, Mannien J, Ott A, Kullberg BJ, de Boer AS, Gyssens IC. Antibiotic prophylaxis and the risk of surgical site infections following total hip arthroplasaty: timely administration is the most important factor. Clin Infect Dis. 2007;44:921–7.
13. Garey KW, Dao T, Chen H, Amrutkar P, Kumar N, Reiter M, et al. Timing of vancomycin prophylaxis for cardiac surgery patients and the risk of surgical site infections. J Antimicrob Chemother. 2006;58:645–50.
14. Blumenthal KG, Ryan EE, Li Y, Lee H, Kuhlen JL, Shenoy ES. The impact of a reported penicillin allergy on surgical site infection risk. Clin Infect Dis. 2017;66(3):329–36.
15. Spencer S, Ipema H, Hartke P, Krueger C, Rodriguez R, Gross A, et al. Intravenous push administration of antibiotics: literature and considerations. Hosp Pharm. 2018;53(3):157–69.
16. Blum S, Cunha CB, Cunha BA. Lack of pharmacokinetic basis of weight-based dosing and intra-operative re-dosing with cefazolin surgical prophylaxis in obese patients: implications for antibiotics stewardship. Surg Infect. 2019;20(6):439–43.
17. Chen X, Brathwaite CE, Barkan A, Hall K, Chu G, Cherasard P, et al. Optimal cefazolin prophylactic dosing for bariatric surgery: no need for higher doses or intraoperative redosing. Obes Surg. 2017;27(3):626–9.
18. Dash CH. Penicillin allergy and the cephalosporins. J Antimicrob Chemother. 1975;1:107–8.
19. Petz LD. Immunologic reactions of humans to cephalosporins. Postgrad Med J. 1971;47s:64–9.
20. Lagace Wiens P, Rubenstein E. Adverse reactions to β-lactam antimicrobials. Expert Opin Drug Saf. 2012;11(3):381–99.
21. Zagursky R, Pichichero ME. Cross-reactivity in β-lactam allergy. J Allergy Clin Immunol Pract. 2018;6(1):72–81.
22. Macy E, Contreras R. Adverse reactions associated with oral and parenteral use of cephalosporins: a retrospective population-based analysis. J Allergy Clin Immunol. 2015;125(3):745–52.
23. Smith JW, Johnson JE, Cluff LE. Studies on the epidemiology of adverse drug reactions: an evaluation of penicillin allergy. NEJM. 1966;274:998–1002.
24. Patterson RA, Stankewicz HA. Penicillin allergy. Treasure Island: StatPearls Publishing; 2019.
25. Macy E, Contreras R. Health care use and serious infection prevalence associated with penicillin "allergy" in hospitalized patuents: a cohort study. J Allergy Clin Immunol. 2014;133(3):790–6.
26. Blumenthal KG, Lu N, Zhabg Y, Li Y, Walensky RP, Choi HK. Risk of methicillin resistant *Staphylococcus aureus* and *Clostridium difficile* in patients with a documented penicillin allergy: population based matched cohort study. BMJ. 2018;361:k2400. https://doi.org/10.1136/bmj.k2400.
27. Lam P, Tarighi P, Elligsen M, Gunaratne K, Nathens A, Tarshis J, et al. Self reported beta lactam allergy and the risk of surgical site infection: a retrospective cohort study. Infect Control Hosp Epidemiol. 2020;41:438–43.
28. Staicu M, Vyles D, Shenoy E, Stone C, Banks T, Alvarez K, et al. Penicillin allergy delabeling: a multidisciplinary opportunity. J Allergy Clin Immunol Pract. 2020;8(9):2858–68.
29. du Plessis T, Walls G, Jordan A, Holland DJ. Implementation of a pharmacist-led penicillin allergy de-labelling service in a public hospital. J Antimicrob Chemother. 2019;74:1438–46. https://doi.org/10.1093/jac/dky575.
30. McDanel DL, Azar AE, Dowden AM, Murray-Bainer S, Noiseux NO, Willenborg M. Screening for beta-lactam allergy in joint arthroplasty patients to improve surgical prophylaxis practice. J Arthroplast. 2017;32(9S):S101–8.

31. Moussa Y, Shuster J, Matte G, Sullivan A, Goldstein RH, Cunningham D, et al. De-labeling of β-lactam allergy reduces intraoperative time and optimizes choice in antibiotic prophylaxis. Surgery. 2018;164:117–23. https://doi.org/10.1016/j.surg.2018.03.004.
32. Kuruvilla M, Sexton M, Wiley Z, Langfitt T, Lynde G, Wolf F. A streamlined approach to optimize perioperative antibiotic prophylaxis in the setting of penicillin allergy labels. J Allergy Clin Immunol. 2020;145(2):AB55.
33. Macy E, Crawford W, Nguyen M, Adams J. Removing the warning not to use cephalosporins in individuals with a penicillin allergy: the effect on cephalosporin allergy and anaphylaxis incidence. J Allergy Clin Immunol. 2020;145(2):AB174.
34. Anderson DJ, Podgorny K, Berrios-Torres SI, Bratzler DW, Dellinger EP, Greene L, et al. Strategies to prevent surgical site infections in acute care hospitals: 2014 update. Infect Control Hosp Epidemiol. 2014;35(6):605–27.
35. Ban KA, Minei JP, Laronga C, Harbrecht BG, Jensen EH, Fry DE, et al. American College of Surgeons and Surgical Infection Society: surgical site infection guidelines, 2016 update. J Am Coll Surg. 2016;224(1):59–74.
36. Allegranzi B, Zayed B, Bischoff P, Kubilay NZ, de Jonge S, de Vries F, et al. New WHO recommendations on intraoperative and postoperative measures for surgical site infection prevention: an evidence-based global perspective. Lancet Infect Dis. 2016;16(12):e288–303.
37. Harbarth S, Samore MH, Lichtenberg D, Carmeli Y. Prolonged antibiotic prophylaxis after cardiovascular surgery and its effect on surgical site infections and antimicrobial resistance. Circulation. 2000;101:2916–21.
38. Cohen ME, Salmasian H, Li J, Liu J, Zachariah P, Wright JD, et al. Surgical antibiotic prophylaxis and risk for postoperative antibiotic-resistant infections. J Am Coll Surg. 2017;225(5):631–8.
39. Kirkwood KA, Gulack BC, Iribarne A, Bowdish ME, Greco G, Mayer ML, et al. A multi-institutional cohort study confirming the risks of *Clostridium difficile* infection associated with prolonged antibiotic prophylaxis. J Thorac Cardiovasc Surg. 2018;155(2):670–768.
40. Branch-Elliman W, O'Brien W, Strymish J, Itani K, Wyatt C, Gupta K. Association of duration and type of surgical prophylaxis with antimicrobial associated adverse events. JAMA Surg. 2019;154(7):590–8.
41. Christensen D, Moschetti W, Brown M, Lucas A, Jevsevar D, Fillingham Y. Perioperative antibiotic prophylaxis: single and 24-hour antibiotic dosages are equally effective at preventing periprosthetic joint infection in total joint arthroplasty. J Arthroplast. 2021;36:S308–13. https://doi.org/10.1016/j.arth.2021.02.037.
42. Thornley P, Evaniew N, Riediger M, Winemaker M, Bhandari M, Ghert M. Postoperative antibiotic prophylaxis in total hip and knee arthroplasty: a systematic review and meta analysis of randomized controlled trials. CMAJ Open. 2015;3(3):338–48.
43. Saraswat MK, Magruder JT, Crawford TC, Gardner JM, Duquaine D, Sussman MS, et al. Preoperative *Staphylococcus aureus* screening and targeted decolonization in cardiac surgery. Ann Thorac Surg. 2017;104(4):1349–56.
44. Sporer SM, Rogers T, Abella L. Methicillin-resistant and methicillin-sensitive *Staphylococcus aureus* screening and decolonization to reduce surgical site infection in elective total joint arthroplasty. J Arthroplast. 2016;31(9S):144–7.
45. Nicolle LE, Gupta K, Bradley SF, Colgan R, DeMuri GP, Drekonja D, et al. Clinical practice guideline for the management of asymptomatic bacteriuria: 2019 update by the Infectious Diseases Society of America. Clin Infect Dis. 2019;68:e83–110. https://doi.org/10.1093/cid/ciy1121.
46. Sousa R, Munoz-Mahamud E, Quayle J, Dias da Costa L, Casals C, Scott P, et al. Is asymptomatic bacteriuria a risk factor for prosthetic joint infection? Clin Infect Dis. 2014;59(1):41–7.
47. Drekonja DM, Zarmbinski B, Johnson JR. Preoperative urine cultures at a veterans affairs medical center. JAMA Intern Med. 2013;173(1):71–2.
48. Bouvet C, Lubbeke A, Bandi C, Pagani L, Stern R, Hoffmeyer P, et al. Is there any benefit in pre-operative urinary analysis before elective total joint replacement? Bone Joint J. 2014;96-B(3):390–4.

49. Cordero-Ampuero J, Gonzalez-Fernandez E, Martinez-Velez D, Esteban J. Are antibiotics necessary in hip arthroplasty with asymptomatic bacteriuria? Seeding risk with/without treatment. Clin Orthop Relat Res. 2013;471(12):3822–9.
50. Gallegos-Salazar O'BW, Strymish JM, Itani K, Branch-Elliman W, Gupta K. Association of screening and treatment for preoperative asymptomatic bacteriuria with positive outcomes among US veterans. JAMA Surg. 2019;154(3):241–8.
51. Gupta K, O'Brien W, Gallegos-Salazar J, Strymish J, Branch-Elliman W. How testing drives treatment in asymptomatic patients: level of pyuria directly predicts probability of antimicrobial prescribing. Clin Infect Dis. 2019;71:614–21. https://doi.org/10.1001/jamasurg.2018.
52. Lamb MJ, Baillie L, Pajak D, Flynn J, Bansal V, Simor A, et al. Elimination of screening urine cultures prior to elective joint arthroplasty. Clin Infect Dis. 2017;64(6):806–9.
53. FDA Drug Safety and Availability 2021. FDA requests withdrawal of bacitracin injection from the market. [Internet] Available from: https://www.fda.gov/drugs/drug-safety-and-availability/fda-requests-withdrawal-bacitracin-injection-market.
54. Damm S. Intraoperative anaphylaxis associated with bacitracin irrigation. Am J Health Syst Pharm. 2011;68(4):323–7.
55. Sacco K, Baumann S, Pongdee T. Intraoperative anaphylaxis secondary to bacitracin irrigation. Ann Allergy Asthma Immunol. 2016;117(5S):S16.
56. Edmiston CE, Leaper D, Spencer M, Truitt K, Litz Fauerbach L, Graham D, et al. Considering a new domain for antimicrobial stewardship: topical antibiotics in the open surgical wound. Am J Infect Control. 2017;45(11):1259–66.
57. Goswami K, Cho JE, Manrique J, Tan T, Higuera C, Della Valle C, et al. Polymyxin and bacitracin in the irrigation solution: there is no role for this practice. Orthop Proc. 2019;101:56. Available from: https://online.boneandjoint.org.uk/doi/abs/10.1302/1358-992X.2019.4.056

Antimicrobial Stewardship in Canine and Feline Veterinary Medicine

6

Claire Fellman and Annie Wayne

Introduction

Antimicrobials are commonly used in veterinary medicine for both prevention and treatment of disease. Prophylactic use is more common in production medicine (i.e., food animal agriculture), whereas treatment of herds and individuals occurs in both production and companion animal medicine. Companion animals encompass animals typically considered pets, such as dogs, cats, and horses. Birds, fish, small mammals, and reptiles, also commonly referred to as exotic species, can also be companion animals. Although antimicrobials are used in all these species, the focus of this chapter is on antimicrobial use in dogs and cats.

Veterinarians treat animals with most of the antimicrobials available to humans, including penicillins, cephalosporins, carbapenems, aminoglycosides, fluoroquinolones, macrolides, lincosamides, rifamycins, tetracyclines, phenicols, nitroimidazoles, and potentiated sulfonamides. Beta-lactams comprise the majority of antimicrobial prescriptions in dogs and cats, with metronidazole, fluoroquinolones, doxycycline, trimethoprim-sulfonamide, and clindamycin also commonly used [1–4]. While there are several veterinary-specific antimicrobials (mostly fluoroquinolones, macrolides, and a few cephalosporins), many of the drugs used in companion animal veterinary medicine are drugs approved for humans and used in an extralabel manner.

Extralabel drug use is defined as the use of an approved drug in a different way than is indicated by the label (different species, indication, dose, frequency, route,

C. Fellman (✉)
Department of Clinical Sciences, Tufts University, North Grafton, MA, USA
e-mail: claire.fellman@tufts.edu

A. Wayne
Massachusetts Veterinary Referral Hospital, Ethos Veterinary Health, Woburn, MA, USA
e-mail: awayne@ethosvet.com

S. Doron, M. Campion (eds.), *Antimicrobial Stewardship in Non-Traditional Settings*, https://doi.org/10.1007/978-3-031-21710-4_6

or duration of treatment). Extralabel drug use is authorized in veterinary medicine by the Animal Medicinal Drug Use Clarification Act of 1994 (AMDUCA) and is common. Some antimicrobials are banned from extralabel use including fluoroquinolones, glycopeptides, and nitroimidazoles in all food animal species, and sulfonamides in lactating dairy cattle. Extralabel cephalosporin use is also restricted but is limited to drugs other than cephapirin, and restrictions are only for cattle, pigs, chickens, and turkeys [5]. These bans are either due to carcinogenicity or the potential for adverse effects in humans (as for nitroimidazoles, sulfas, and chloramphenicol) or to reduce the development of antimicrobial resistance (AMR) (as for fluoroquinolones, cephalosporins, and glycopeptides). There are currently no antimicrobial use restrictions for companion animal prescribing in the USA.

By mass, the majority of antimicrobials prescribed in veterinary medicine are in production medicine, and there is a concerted effort to monitor the development of AMR in the animals and animal products entering the human food chain. Current research efforts are trying to establish the duration needed for treatment of various diseases in production animals to reduce antimicrobial use, and recent legislation banned antimicrobial use for production purposes (feed efficiency and growth promotion) and brought all remaining medically important antimicrobials administered in feed under veterinary oversight by requiring a Veterinary Feed Directive (VFD) from a licensed veterinarian. Further discussion of the advances in antimicrobial stewardship in production medicine is beyond the scope of this chapter. However, it is noteworthy that these efforts contrast strongly with the nascent status of stewardship efforts in companion animals in the USA, which largely consist of recommendations and a few guidelines, with little national or government support for stewardship program implementation.

Veterinary medicine is growing in complexity and in the availability and demand for specialty care. There are presently more than 20 American Veterinary Medical Association-recognized veterinary specialties and specialty organizations mirroring human medicine, including ophthalmology, cardiology, neurology, surgery, internal medicine, dermatology, nutrition, and theriogenology. Clients seeking veterinary care visit general practitioners with a Doctor of Veterinary Medicine (DVM) degree or seek care at referral hospitals with boarded diplomates of the various specialties. Efforts to improve antimicrobial prescribing must therefore incorporate both the common infections encountered in general practice such as urinary tract, skin, and upper respiratory infections and the more complicated cases treated by specialists.

The reader may ask, why should we care about antimicrobial prescribing in companion animals? Although most veterinary antimicrobial use occurs in production animals, companion animals are in much closer contact with their human owners than production animals, and both increasing levels of AMR and resistant organism transmission between humans and companion animals have been identified in recent studies. The first portion of this chapter examines evidence for resistance transmission and shared flora between pets and their owners. Subsequent sections evaluate the information available in existing guidelines, the present status of stewardship efforts for dogs and cats, and future areas for research and improved antimicrobial use.

One Health is an approach that acknowledges that the health of humans is inextricably connected to that of animals and the shared environment. As the global crisis of AMR continues, it is imperative to employ a One Health approach and that all antimicrobial use be evaluated and optimized.

Resistance Prevalence and Shared Flora Between Pets and their Owners

Pets and their owners are often in extremely close contact. While some dogs and cats live exclusively outdoors, many pets sleep in bed with their owners and some share food or drink with their owners and/or have mouth to mouth contact. Facilitated by close contact, infections can be transmitted from pets to humans through bites or scratches and arthropod vectors (i.e., ticks, fleas). Given the long-term, intimate nature of human–animal relationships, it is unsurprising that such contact can lead to shared bacterial flora. Zoonotic infections are those that are transmitted between animals and humans, or vice versa. While some bacterial pathogens may be transmitted between humans and animals and cause disease (such as methicillin-resistant *Staphylococcus aureus,* MRSA), resistance features or commensal and colonizing bacterial species can also be shared between pets and their owners.

One of the earliest recognitions that pets could play a role in AMR transmission was the 1988 report of a cat living in a human hospital ward that was colonized with MRSA and possibly spreading disease to hospitalized geriatric patients. In this report, removal of the cat led to resolution of the outbreak [6]. There are many other reports of shared *Staphylococcus* spp. including suspected transmission of MRSA from a horse to its owner [7], transmission of MRSA from pigs to a veterinarian and then from the veterinarian to their dog [8], MRSA transmission from an individual with a decubitus ulcer to their dog [8], and shared carriage of closely related *Staphylococcus haemolyticus* between a cat and its owner [9]. Most studies recognize MRSA transmission generally occurs from humans to animals (rather than the opposite), although horses can become colonized and are frequent carriers of the organism.

Resistant Enterobacteriaceae are also common health concerns, and there is increasing research and study into the potential for pets to act as reservoirs for resistant organisms. The zoonotic potential of nontyphoidal *Salmonella* is well recognized, and ampC beta-lactamase-producing strains are frequently identified from animals including dogs, cats, and horses [10]. *Escherichia coli* is also commonly studied with one report finding similar virulence gene and antimicrobial resistance profiles among fluoroquinolone-resistant isolates from dogs and humans suggesting the potential for organism exchange between the two species [11]. Another study documented clonally related human extended-spectrum beta-lactamase (ESBL)-producing *E. coli* strains in companion animals from various European countries, again suggesting the possibility of inter-species transmission [12]. Two studies have assessed household carriage of *E. coli* as a risk factor for urinary tract infection and found frequent sharing of clones between household members including humans and dogs suggesting either transmission or a common source, such as food [13, 14]. A study of pregnant women found more frequent vaginal colonization with *E. coli* in women with cats and dogs, with higher antimicrobial prescription rates in pet-owning women, and higher self-reported urinary tract infection rates in women with cats [15].

Many studies have evaluated the prevalence of resistance features like ESBL production in canine and feline bacterial isolates resulting in identification of antimicrobial resistance features shared with human isolates. One Chinese study found an increasing prevalence of multidrug-resistant and ESBL-producing *E. coli* in pets over the 5-year study period (2012–2017) that correlated with beta-lactam usage. Study authors warned of the potential for spread of resistant organisms to humans and the need to pursue stewardship interventions in companion animals [16]. Importantly, carbapenemase-producing *E. coli* have been documented from infections in dogs and cats in the USA [17] and dogs in Europe [18]. A Chinese study found carbapenemase-producing *E. coli* in dogs and other backyard animals including pigs and chickens and suggested inter-host transmission between humans and animals [19]. Carbapenem use in veterinary medicine is restricted in some European countries, but as mentioned above, no current use regulations in companion animals exist in the USA for this class or any other antimicrobials.

Transmission of other organisms between animals and humans such as *Campylobacter* and *Leptospirosis* is recognized, and acquisition of concerning AMR in companion animals has also been documented for shared pathogens like *Pseudomonas* and *Acinetobacter*. While it is challenging to prove direct transmission of organisms, and even harder to prove transmission of genetic resistance elements, there is overwhelming evidence to support this likelihood. AMR complicates treating individual animals and given the similarity of pathogens, shared resistance elements, and evidence for inter-species transmission and household sharing of bacteria, stewardship interventions and improved antimicrobial use are critical for companion animal veterinary medicine.

Antimicrobial Stewardship Principles and Perceptions in Companion Animal Medicine

The American Veterinary Medical Association (AVMA) standing Committee on Antimicrobials, formed in 2016, defines antimicrobial stewardship for veterinarians as "the actions veterinarians take individually and as a profession to preserve the effectiveness and availability of antimicrobial drugs through conscientious oversight and responsible medical decision-making while safeguarding animal, public, and environmental health" and lays out the following core principles for all veterinary professionals, including small animal practitioners [20]:

1. Commit to stewardship
2. Advocate for a system of care to prevent common diseases

3. Select and use antimicrobial drugs judiciously
4. Evaluate antimicrobial drug use practices
5. Educate and build expertise

In addition, the committee has published a position statement on "Judicious Therapeutic Use of Antimicrobials" [21]. The judicious use principles outlined in the document include recommendations for disease prevention, endorsement of prescribing of antimicrobials by veterinarians in compliance with a valid veterinarian–client–patient relationship, use of microbial culture and susceptibility testing whenever possible, and prescribing antimicrobials only for appropriate infections and to minimize treatment length [21]. Importantly, the veterinarian's oath includes a mandate not only to protect animal health but public health as well. The AVMA has endorsed several additional publications and guidelines for antimicrobial use in animal species beyond the scope of this chapter including aquatic animals, poultry, cattle, horses, and swine [22–25].

In 2013, the AVMA created the *Task Force on Antimicrobial Stewardship in Companion Animal Practice*, whose stated goal is to "consider the emerging impact of multidrug-resistant organisms in companion animal practice and design approaches to address this challenge" [26].

The task force outlines strategies for companion animal veterinary medicine stakeholders to improve stewardship efforts and provides educational materials for veterinarians and clients, including posters with "do's and don'ts" for prescribing antimicrobials in dogs and cats. The core elements outlined in the 2016 report from the task force are [27]:

1. Each veterinary practice should commit to the reduction of antimicrobial resistance by creating and adopting a clinic-specific plan in which everyone contributes to effective antimicrobial stewardship.
2. Antimicrobial stewardship in a practice begins with the appointment of a principal person to lead the program and to support the practice stewardship goals with all available resources.
3. Practices should address issues of antimicrobial resistance by taking a proactive approach to antimicrobial use and infection control using accepted guidelines, recommendations, and expertise.
4. Understanding what is (and what is not) working requires self-assessment and monitoring with a goal of continual improvement.

In addition, the task force recommends the use of the "antimicrobial time-out" re-evaluation 48–72 h after initiation of antimicrobial therapy and the creation of systems for tracking and monitoring measurable outcomes for stewardship interventions both in clinics and more broadly for companion animal veterinary medicine. The recommendations of the AVMA's committees on stewardship provide excellent goals for the profession. To date, however, few resources (whether personnel, research support, or financial incentives) have been made available for the implementation of these recommendations.

Internationally, in Canada, Australia, and some European nations, similar efforts to provide guidelines for prudent use have been published recently. There is a body of literature examining antimicrobial use patterns and prescription rates [1, 3, 4, 28–39], and studies that show introduction of stewardship tools and education reduce overall use [40–43]. Although large-scale studies of antimicrobial prescribing have been reported using data from medical records and pet insurance companies, studies are often limited by a lack of precise information regarding treatment duration and dose. Member states of the European Union will be required to report antimicrobial use for all animals (including companion animals) by 2030 [44]. There is no reporting mandate for antimicrobial use in animals in the USA, but similar measures and infrastructure development would greatly facilitate stewardship programming and assessing the effects of interventions.

Several recent publications have examined small animal veterinarians and pet owners' knowledge and perceptions of antimicrobial stewardship, which have the potential to inform next steps for stewardship interventions and efforts. Based on current literature, small animal practitioners are concerned about antimicrobial resistance, and there is a desire for more support for practicing better stewardship in the clinic. A 2015 survey of 71 faculty at a teaching hospital (large and small animal veterinarians) found 59% were very concerned about antimicrobial resistance and 88% felt antimicrobials were overprescribed [45]. A 2015 survey of 262 general practitioner veterinarians in the USA found 82% were strongly or somewhat concerned about antimicrobial resistance in dogs and cats, but 88% of respondents were unaware of available guidelines. 77% strongly or somewhat agreed that more guidance on which antimicrobial to use and 83% strongly or somewhat agreed that more guidance on duration of treatment would be helpful. In this survey, 84% strongly or somewhat agreed that cost of culture and susceptibility testing was a barrier to obtaining this information [46]. In contrast, a 2019 survey of 25 pet owners found that most participants generally knew what appropriate antimicrobial use was, but few were concerned with antimicrobial resistance [47]. A unique 2018 study in which 21 pet owners and 16 veterinarians were interviewed in the United Kingdom showed that veterinarians may perceive pressure from pet owners to prescribe antimicrobials but that pet owners generally felt antimicrobials were overprescribed and would trust a veterinarian's decision not to prescribe if inappropriate. This study found that pet owners' knowledge of antimicrobial resistance was low [48]. A larger study of North American dog owners similarly found low understanding of antimicrobial resistance with most dog owners preferring to use drugs considered very important to human medicine in their pets, and conjoint analysis showing cost to be the most important factor for antimicrobial preference by dog owners [49]. These studies highlight the importance of antimicrobial stewardship education for both veterinarians and pet owners.

Current Status of Antimicrobial Stewardship Efforts and Available Guidelines in the USA

Although national organizations such as the AVMA recommend it, there is no requirement for individual veterinary small animal hospitals to have infection control or antimicrobial stewardship teams or restrictions in prescribing antimicrobials. There are extensive data in the human literature showing the impact of various stewardship efforts on antimicrobial use patterns. Similarly, a 2006 study in veterinary hospitals found that guidelines in a small animal veterinary teaching hospital decreased antimicrobial prescriptions per hospital admission over the study period. However, at the time of the study, only 3 out of 21 teaching hospitals in the USA and Canada had specific policies around prescribing for higher tier antimicrobials such as amikacin, imipenem, and vancomycin [50]. Unfortunately, veterinary hospital-level antimicrobial prescribing recommendations remain limited. In an informal survey regarding carbapenem prescribing of 127 diplomates via the listservs for the American College of Veterinary Emergency and Critical Care and the American College of Veterinary Internal Medicine, 13% of responders said their practice had some sort of review process for use of carbapenems, with only 6% of responders describing official committees or a sign-off process. When respondents were asked to describe the policies and guidelines, most described a culture-driven process and none referenced official written guidelines at their facilities (unpublished data, Wayne and Fellman 2017). Additionally, pharmacists and microbiologists, often key personnel in human hospital stewardship teams, are rarely employed by and even more rarely involved in stewardship efforts in small animal veterinary settings. Infectious disease as a formal specialty in veterinary medicine does not exist, and there are only a select few small animal veterinarians with expertise in infectious disease gained from clinical experience and research in the USA. Almost no primary care veterinarians have resources to invest in stewardship teams. To date, there has been limited outside funding and few incentives for veterinary hospitals to implement stewardship or surveillance programs, although concern around antimicrobial resistance and interest in extending stewardship efforts to companion animal medicine is growing. In the last few years, several US veterinary academic institutions have devoted resources and funds to implement stewardship programs and increase AMR awareness in veterinary trainees and graduates, and a few grants and contracts focusing on small animal stewardship and antimicrobial use tracking have been awarded by the Food and Drug Administration and Centers for Disease Control and Prevention.

In addition to the lack of resources, other challenges exist for implementing companion animal antimicrobial stewardship programs. Compared to the existing human literature, few guidelines are available to direct small animal antimicrobial prescribing, and the published evidence to support these veterinary guidelines is relatively scant. Laboratory and microbiology support for small animal veterinarians is also limited and variable. Most hospitals are not large enough to support clinic-specific antibiograms, and there are few regions with coordinated efforts to provide regional antibiograms, although increasingly, there is discussion and education around antibiogram development and use [51, 52]. Barriers to implementation of antibiograms more widely include lack of expertise, standardization, and funding. A 2015 survey of 208 veterinarians in Washington state found that only 16% of respondents had access to a clinic-specific antibiogram [53]. The Clinical and Laboratory Standards Institute (CLSI) has published veterinary standards for testing and interpretation for many antimicrobials used in dogs and cats, but veterinary-specific breakpoints are not available for all drugs and are presently limited to certain bacteria in certain animal species (cats, dogs, horses, cattle, pigs, fish) [54]. Human breakpoints may be used by the diagnostic laboratory when veterinary interpretations are not available, and ongoing efforts to establish veterinary breakpoints have shown in some cases the human breakpoints are appropriate, and in others veterinary values are considerably different and historical interpretations using human values have misidentified antimicrobial susceptibility/resistance [55]. An additional challenge for appropriate antimicrobial administration is that while some drugs have well-established effective dosages, other less commonly used antimicrobials rely at times on minimal pharmacokinetic data that are often extrapolated across veterinary species to establish dosages.

In veterinary medicine, unlike in human medicine, every diagnostic test is a direct cost to owners. This provides an opportunity for consideration by practitioners and education to owners regarding the utility of each test but can also prevent important diagnostic information from being collected and encourages empiric use, which may hinder stewardship efforts. It is also harder to obtain certain cultures from dogs and cats than humans. For example, the respiratory tract is uncommonly sampled despite the relatively high prevalence of pneumonia in dogs because sputum cultures are not possible and samples must be collected using tracheal wash or bronchoalveolar lavage requiring sedation or anesthesia. These special considerations need to be incorporated into guidelines and stewardship efforts in dogs and cats.

Table 6.1 Recommended antimicrobial stewardship and infection control guidelines

Title	Reference
American Association of Feline Practitioners (AAFP)/American Animal Hospital Association (AAHA) 2022 Antimicrobial Stewardship Guidelines	[56]
American College of Veterinary Internal Medicine (ACVIM) Consensus Statement on Antimicrobial Drug Use in Veterinary Medicine	[57]
American College of Veterinary Internal Medicine (ACVIM) Consensus Statement on Therapeutic Antimicrobial Use in Animals and Antimicrobial Resistance	[58]
Infection Prevention and Control Best Practice Guidelines for Small Animal Veterinary Clinics	[59]
Handbook of Antibiotic Stewardship in Companion Animal Veterinary Settings	[60]

The guidelines in Table 6.1 outline general principles and mirror recommendations in human medicine such as adhering to best practices for infection control, developing surveillance and monitoring systems, using culture and susceptibility testing to guide therapeutic use whenever possible, limiting the empiric use of antimicrobials, using the narrowest spectrum and shortest course considered effective, de-escalating as indicated, and using antimicrobials considered critically important in human medicine sparingly and only after consultation with appropriate specialists.

Maintaining the availability of effective antimicrobials for animals is a priority for veterinary organizations, and the AVMA encourages promoting appropriate use within the profession. A joint statement from the AVMA, the Federation of Veterinarians of Europe, and the Canadian Veterinary Medical Association advocates that when considered, "Regulatory proposals and actions should be specific, transparent, and based on scientific risk analysis" [61].

The leading organization writing clinical treatment guidelines for companion animal veterinary medicine, the International Society for Companion Animal Infectious Diseases (ISCAID), was founded in 2006 and has used working groups comprised of veterinary and microbiology experts to draft specific guidelines for urinary tract infections (2011, updated 2019), canine superficial bacterial folliculitis (2014), and respiratory tract disease in dogs and cats (2017). However, it is important to note that many of the recommendations from the working groups are largely based on expert opinion, as supporting data for recommendations, particularly in length of treatment with antimicrobials, are lacking in the veterinary literature. The following are some of the key points from each of the guidelines.

Urinary Tract Infections (Original Publication in 2011, Updated in 2019) [62, 63]

Diagnosis of sporadic urinary tract infections in dogs and cats is based on clinical signs plus supporting clinical pathologic evidence of disease including hematuria, pyuria, bacteriuria, and bacterial culture results. Aerobic bacterial cultures are recommended in all cases, but the committee feels empiric treatment without a culture in dogs is reasonable if there are no other concurrent diseases and no recent history of antimicrobial therapy. Cats are prone to lower urinary tract syndrome (LUTS), a sterile cystitis, so all cats with lower urinary tract signs that do not respond to empiric therapy for LUTS (pain medication and stress reduction) should have a culture performed before antimicrobial therapy is prescribed. Ideally, samples for culture in small animals are collected via cystocentesis. Recommended duration of therapy for sporadic uncomplicated urinary tract infections is 3–5 days, but supporting evidence on duration of treatment in dogs and cats is very limited [64–66]. First-line antimicrobials for dogs include amoxicillin +/− clavulanate or trimethoprim-sulfonamide and consider nitrofurantoin, fluoroquinolones, or third-generation cephalosporins if first-line therapy is not appropriate based on culture and sensitivity testing or patient factors. The guidelines contain additional information regarding complicated infections, pyelonephritis, and catheter-associated infections.

Canine Superficial Bacterial Folliculitis (2014) [67]

Dogs are prone to pyoderma, most commonly associated with allergic disease states. The most common cause of superficial bacterial folliculitis is *Staphylococcus pseudintermedius*, which is a commensal pathogen. Diagnosis of superficial bacterial folliculitis is based on clinical signs and hallmark erythematous papules and pustules associated with hair follicles. Cytology and/or culture to diagnose co-infections such as yeast, dermatophytes, and other bacteria is recommended. Treatment includes systemic antimicrobial therapy in addition to topical therapies including shampoos, sprays, rinses, and lotions and modifying diet or environment to minimize allergic components of disease wherever possible. Recommended first-line empiric antimicrobial therapy is clindamycin or lincomycin, amoxicillin-clavulanate, or first-generation cephalosporins. Culture is strongly recommended if empiric first-line therapy fails and/or there is:

- Less than 50% reduction in lesions within 2 weeks of empiric systemic antimicrobial therapy
- Emergence of new lesions during the course of antimicrobial therapy
- Residual lesions after 6 weeks of systemic antimicrobial therapy
- Intracellular rod-shaped bacteria seen on cytology
- Prior history of multidrug-resistant infections in the household

Antimicrobial Use Guidelines for Treatment of Respiratory Tract Disease in Dogs and Cats (2017) [68]

Both dogs and cats, similarly to humans, are commonly exposed to viral upper respiratory tract infections, and treatment with antimicrobials is discouraged. If the dog or cat displays clinical signs parallel to the human cold (sneezing, nasal and/or ocular discharge, lethargy, decreased appetite) but is still eating and drinking and does not have tachypnea or increased respiratory effort, observation for 10 days is recommended before empiric antimicrobial therapy should be considered. If signs persist for longer than 10 days, mucopurulent discharge is noted, or respiration affected, antimicrobial therapy and additional diagnostics are recommended.

Pneumonia in cats is uncommon, and other concurrent diseases should be considered and ruled out with diagnostics. In dogs, pneumonia secondary to viral infections or due to aspiration is relatively common. Clinical diagnosis is usually made based on physical exam and thoracic radiographic evidence of pulmonary infiltrates.

Cytology and culture of the airways are recommended and ideal, but in dogs and cats, airway sampling requires at least sedation (direct transtracheal samples can be performed in large breed dogs) or general anesthesia for endotracheal sampling or bronchoalveolar lavage. The cost and ability to perform these tests can be a deterrent for clinicians and pet owners, particularly in the primary care setting. Current dogma in veterinary medicine is to treat pneumonia for 4–6 weeks [68] and/or 1–2 weeks beyond radiographic resolution of pulmonary infiltrates, but this recommendation is not evidence-based [69] and is clearly far greater than the duration of therapy used for the same condition in human medicine, suggesting an urgent need for research. There is limited evidence to suggest that shorter courses in uncomplicated canine pneumonia are not inferior [70–72], and the working group recommends re-evaluation of clinical signs, hematology lab results, and radiographs 10–14 days after initiating therapy. First-line therapy for community-acquired pneumonia with relatively mild clinical signs and high suspicion of *Bordetella* or *Mycoplasma* co-infection is doxycycline. For acute aspiration pneumonia, amoxicillin, amoxicillin-clavulanate, or a first-generation cephalosporin is recommended. For dogs and cats with sepsis caused by pneumonia, a fluoroquinolone plus amoxicillin/ampicillin or clindamycin is recommended. These guidelines also include recommendations for treatment of pyothorax and complicated respiratory infections.

Consensus statements provide additional guidance for the diagnosis and treatment of Lyme disease [73], leptospirosis [74], methicillin-resistant staphylococcal infections [75], and ehrlichial diseases in dogs and cats [76], but are beyond the scope of this chapter.

Summary of Issues

Despite the body of evidence (in human healthcare settings and corroborated in veterinary settings) that indicates the effectiveness of surveillance, guidelines, and local stewardship efforts, there are relatively few resources currently devoted to implementing such efforts in small animal veterinary settings. Even at tertiary care veterinary teaching hospitals, specific programs and financial investment in infection control and antimicrobial stewardship are highly variable. Primary veterinary care for dogs and cats is mostly provided by individual practices, sometimes with only one or two veterinarians who often are also the owners of the practice. In the last 10 years, there has been a trend toward corporatization of veterinary medicine. However, the extent to which the corporation gives oversight or guidelines for practitioners in their day-to-day practice varies widely. Stewardship at the primary care level requires reaching individual clinicians and providing effective tools that can be adapted to a variety of types of clinics and practitioners. Although the framework and underlying principles for stewardship in veterinary medicine mirror human medicine, the fact that pet owners pay for all diagnostics and treatments for their pets out of pocket is a unique aspect that requires specific study and interventions.

Strategies for the Future

The momentum and desire for improved companion animal stewardship is growing as the One Health approach gains traction and the importance of antimicrobial use in dogs and cats is increasingly recognized. Although the number of available guidelines and educational materials is increasing, there is limited evidence assessing the impact of existing initiatives, largely due to our current inability to measure companion animal antimicrobial prescriptions. Efforts to quantitate prescribing and establish benchmarks for appropriate use are developing with a variety of use metrics described.

Researchers in the Netherlands applied a metric previously only used to quantify food animal antimicrobial use in Europe called the defined daily doses animal (DDDA) to quantitate prescribing at Dutch companion animal hospitals. The DDDA evaluates the amount of antimicrobial dispensed by the companion animal hospital divided by the number of animals (dogs, cats, and rabbits) seen that could be prescribed each drug [29, 77]. Although precise prescribing data for each individual are not available, this metric is reproducible and relatively easy to calculate, and a recent study using DDDA showed decreasing antimicrobial use in clinics that was enhanced by the implementation of a veterinary antimicrobial stewardship program [40]. Another study which compared animal daily doses (ADD), a dose-based metric, to days of therapy (DOT) in dogs, large animal patients and dairy herds found poor correlation between the measurements, suggesting additional work to determine the best approach to measuring overall antimicrobial use in veterinary patients is needed and may be different across species and settings [78]. Other efforts to measure antimicrobial prescribing exist using a variety of strategies including point-prevalence and other cross-sectional surveys, retrospective longitudinal studies, and prospective data capture from electronic medical records. Recent studies of Australian medical records reported natural language processing algorithms to mine free text that allowed measurement of antimicrobial prescribing frequency in over four million consults [3, 30]. Large-scale monitoring of prescription rates has also been performed in the United Kingdom through various medical records databases [1, 4], and a recent USA study evaluated antimicrobial prescription rates by service at three academic veterinary hospitals over a 6-year period [2]. As antimicrobial use reporting increases for companion animal veterinary practices, risk adjustment for factors increasing the likelihood of antimicrobial prescription will be necessary to accurately benchmark and compare rates between different facilities.

Other ongoing research focuses on clarifying transmission risks between animals and humans and identifying AMR patterns and resistance genes in companion animals. Still other researchers are evaluating social science interventions to change prescribing patterns and better understand veterinary practitioner and owner perceptions of antimicrobial stewardship.

The path forward is multidisciplinary and involves a variety of approaches to successfully improve antimicrobial use in companion animals. Continued assessment of resistance mechanisms will allow the creation of strategies to mitigate resistance development. Essential to implementing these strategies are the abilities to successfully change prescriber habits and subsequently measure the effect of interventions using both ongoing bacterial susceptibility data and antimicrobial prescription data.

International coordination and collaboration through meetings and targeted funding mechanisms are vital to the continued growth of this field. Taking advantage of the stewardship progress in human medicine, a One Health approach in which companion animal veterinarians partner with physicians and pharmacists can provide proven strategies that can be implemented and adapted. Companion animal antimicrobial stewardship is challenging because of the diversity of clinic settings, variable owner willingness and ability to pursue recommended diagnostic testing, and limited evidence to support best prescribing practices for common conditions. Continued research will improve the evidence base for treatment guidelines, but the other factors will likely remain. Successful antimicrobial stewardship strategies will need to thoughtfully address these veterinary challenges.

References

1. Buckland EL, O'Neill D, Summers J, Mateus A, Church D, Redmond L, et al. Characterisation of antimicrobial usage in cats and dogs attending UK primary care companion animal veterinary practices. Vet Rec. 2016;179(19):489.
2. Goggs R, Menard JM, Altier C, Cummings KJ, Jacob ME, Lalonde-Paul DF, et al. Patterns of antimicrobial drug use in veterinary primary care and specialty practice: a 6-year multi-institution study. J Vet Intern Med. 2021;35:1496–508.
3. Hur BA, Hardefeldt LY, Verspoor KM, Baldwin T, Gilkerson JR. Describing the antimicrobial usage patterns of companion animal veterinary practices; free text analysis of more than 4.4 million consultation records. PLoS One. 2020;15(3):e0230049.
4. Singleton DA, Sanchez-Vizcaino F, Dawson S, Jones PH, Noble PJM, Pinchbeck GL, et al. Patterns of antimicrobial agent prescription in a sentinel population of canine and feline veterinary practices in the United Kingdom. Vet J. 2017;224:18–24.
5. 21CFR530.41. Drugs prohibited for extralabel use in animals.
6. Scott GM, Thomson R, Malone-Lee J, Ridgway GL. Cross-infection between animals and man: possible feline transmission of Staphylococcus aureus infection in humans? J Hosp Infect. 1988;12(1):29–34.
7. van Duijkeren E, Ten Horn L, Wagenaar JA, de Bruijn M, Laarhoven L, Verstappen K, et al. Suspected horse-to-human transmission of MRSA ST398. Emerg Infect Dis. 2011;17(6):1137–9.
8. Nienhoff U, Kadlec K, Chaberny IF, Verspohl J, Gerlach GF, Schwarz S, et al. Transmission of methicillin-resistant Staphylococcus aureus strains between humans and dogs: two case reports. J Antimicrob Chemother. 2009;64(3):660–2.
9. Deng F, Wang H, Liao Y, Li J, Fessler AT, Michael GB, et al. Detection and genetic environment of pleuromutilin-lincosamide-streptogramin a resistance genes in staphylococci isolated from pets. Front Microbiol. 2017;8:234.
10. Frye JG, Fedorka-Cray PJ. Prevalence, distribution and characterisation of ceftiofur resistance in salmonella enterica isolated from animals in the USA from 1999 to 2003. Int J Antimicrob Agents. 2007;30(2):134–42.
11. Platell JL, Cobbold RN, Johnson JR, Heisig A, Heisig P, Clabots C, et al. Commonality among fluoroquinolone-resistant sequence type ST131 extraintestinal Escherichia coli iso-

lates from humans and companion animals in Australia. Antimicrob Agents Chemother. 2011;55(8):3782–7.
12. Ewers C, Grobbel M, Stamm I, Kopp PA, Diehl I, Semmler T, et al. Emergence of human pandemic O25:H4-ST131 CTX-M-15 extended-spectrum-beta-lactamase-producing Escherichia coli among companion animals. J Antimicrob Chemother. 2010;65(4):651–60.
13. Damborg P, Nielsen SS, Guardabassi L. Escherichia coli shedding patterns in humans and dogs: insights into within-household transmission of phylotypes associated with urinary tract infections. Epidemiol Infect. 2009;137(10):1457–64.
14. Johnson JR, Owens K, Gajewski A, Clabots C. Escherichia coli colonization patterns among human household members and pets, with attention to acute urinary tract infection. J Infect Dis. 2008;197(2):218–24.
15. Stokholm J, Schjorring S, Pedersen L, Bischoff AL, Folsgaard N, Carson CG, et al. Living with cat and dog increases vaginal colonization with E. coli in pregnant women. PLoS One. 2012;7(9):e46226.
16. Chen Y, Liu Z, Zhang Y, Zhang Z, Lei L, Xia Z. Increasing prevalence of ESBL-producing multidrug resistance Escherichia coli from diseased pets in Beijing, China from 2012 to 2017. Front Microbiol. 2019;10:2852.
17. Shaheen BW, Nayak R, Boothe DM. Emergence of a New Delhi metallo-β-lactamase (NDM-1)-encoding gene in clinical Escherichia coli isolates recovered from companion animals in the United States. Antimicrob Agents Chemother. 2013;57(6):2902–3.
18. Stolle I, Prenger-Berninghoff E, Stamm I, Scheufen S, Hassdenteufel E, Guenther S, et al. Emergence of OXA-48 carbapenemase-producing Escherichia coli and Klebsiella pneumoniae in dogs. J Antimicrob Chemother. 2013;68(12):2802–8.
19. Li J, Bi Z, Ma S, Chen B, Cai C, He J, et al. Inter-host transmission of carbapenemase-producing Escherichia coli among humans and backyard animals. Environ Health Perspect. 2019;127(10):107009.
20. Antimicrobial stewardship definition and core principles. American Veterinary Medical Association. Available from: https://www.avma.org/policies/antimicrobial-stewardship-definition-and-core-principles.
21. Judicious therapeutic use of antimicrobials. American Veterinary Medical Association. Available from: https://www.avma.org/sites/default/files/resources/2014S_Resolution5_Attch.pdf.
22. Judicious therapeutic use of antimicrobials in aquatic animal medicine. Available from: https://www.avma.org/resources-tools/avma-policies/judicious-therapeutic-use-antimicrobials-aquatic-animal-medicine.
23. AABP. Judicious therapeutic use of antimicrobials in cattle 2019. Available from: https://aabp.org/Resources/AABP_Guidelines/AABP_Judicious_Therapeutic_Use_of_Antimicrobials_Guideline%20-%202019.pdf.
24. AAEP. Judicious use of antimicrobials 2020. Available from: https://aaep.org/guidelines/aaep-ethical-and-professional-guidelines/aaep-position-statements/veterinary-practice.
25. AASV. Basic guidelines of judicious therapeutic use of antimicrobials in swine 2020. Available from: https://www.aasv.org/documents/JUG.php.
26. Bender JB, Barlam TF, Glore RP, Gumley N, Grayzel SE, Hoang C, et al. The AVMA task force for antimicrobial stewardship in companion animal practice responds. J Am Vet Med Assoc. 2015;246(7):727–8.
27. Task Force on Antimicrobial Stewardship in Companion Animal Practice Activities: August 2013–December 2015. American Veterinary Medical Association. Available from: https://www.avma.org/KB/Resources/Reports/Documents/TFASCAP_Report.pdf.
28. Hardefeldt LY, Selinger J, Stevenson MA, Gilkerson JR, Crabb H, Billman-Jacobe H, et al. Population wide assessment of antimicrobial use in dogs and cats using a novel data source – a cohort study using pet insurance data. Vet Microbiol. 2018;225:34–9.
29. Hopman NEM, van Dijk MAM, Broens EM, Wagenaar JA, Heederik DJJ, van Geijlswijk IM. Quantifying antimicrobial use in Dutch companion animals. Front Vet Sci. 2019;6:158.

30. Hur B, Hardefeldt LY, Verspoor K, Baldwin T, Gilkerson JR. Using natural language processing and VetCompass to understand antimicrobial usage patterns in Australia. Aust Vet J. 2019;97(8):298–300.
31. Joosten P, Ceccarelli D, Odent E, Sarrazin S, Graveland H, Van Gompel L, et al. Antimicrobial usage and resistance in companion animals: a cross-sectional study in three European countries. Antibiotics. 2020;9(2):87.
32. Lutz B, Lehner C, Schmitt K, Willi B, Schupbach G, Mevissen M, et al. Antimicrobial prescriptions and adherence to prudent use guidelines for selected canine diseases in Switzerland in 2016. Vet Rec Open. 2020;7(1):e000370.
33. Mateus A, Brodbelt DC, Barber N, Stark KD. Antimicrobial usage in dogs and cats in first opinion veterinary practices in the UK. J Small Anim Pract. 2011;52(10):515–21.
34. Radford AD, Noble PJ, Coyne KP, Gaskell RM, Jones PH, Bryan JGE, et al. Antibacterial prescribing patterns in small animal veterinary practice identified via SAVSNET: the small animal veterinary surveillance network. Vet Rec. 2011;169(12):310–U91.
35. Rantala M, Holso K, Lillas A, Huovinen P, Kaartinen L. Survey of condition-based prescribing of antimicrobial drugs for dogs at a veterinary teaching hospital. Vet Rec. 2004;155(9):259–62.
36. Schmitt K, Lehner C, Schuller S, Schupbach-Regula G, Mevissen M, Peter R, et al. Antimicrobial use for selected diseases in cats in Switzerland. BMC Vet Res. 2019;15(1):94.
37. Mendez M, Moreno MA. Quantifying antimicrobial exposure in dogs from a longitudinal study. Front Vet Sci. 2020;7:545.
38. Escher M, Vanni M, Intorre L, Caprioli A, Tognetti R, Scavia G. Use of antimicrobials in companion animal practice: a retrospective study in a veterinary teaching hospital in Italy. J Antimicrob Chemother. 2011;66(4):920–7.
39. Gomez-Poveda B, Moreno MA. Antimicrobial prescriptions for dogs in the capital of Spain. Front Vet Sci. 2018;5:309.
40. Hopman NEM, Portengen L, Hulscher M, Heederik DJJ, Verheij TJM, Wagenaar JA, et al. Implementation and evaluation of an antimicrobial stewardship programme in companion animal clinics: a stepped-wedge design intervention study. PLoS One. 2019;14(11):e0225124.
41. Hubbuch A, Schmitt K, Lehner C, Hartnack S, Schuller S, Schupbach-Regula G, et al. Antimicrobial prescriptions in cats in Switzerland before and after the introduction of an online antimicrobial stewardship tool. BMC Vet Res. 2020;16(1):229.
42. Lehner C, Hubbuch A, Schmitt K, Schuepbach-Regula G, Willi B, Mevissen M, et al. Effect of antimicrobial stewardship on antimicrobial prescriptions for selected diseases of dogs in Switzerland. J Vet Intern Med. 2020;34(6):2418–31.
43. Singleton DA, Rayner A, Brant B, Smyth S, Noble PM, Radford AD, et al. A randomised controlled trial to reduce highest priority critically important antimicrobial prescription in companion animals. Nat Commun. 2021;12(1):1593.
44. Regulation (EU) 2019/6 of the European Parliament and of the Council of 11 December 2018 on Veterinary Medicinal Products and Repealing Directive 2001/82/EC. European Parliament and The Council; 2019.
45. Jacob ME, Hoppin JA, Steers N, Davis JL, Davidson G, Hansen B, et al. Opinions of clinical veterinarians at a US veterinary teaching hospital regarding antimicrobial use and antimicrobial-resistant infections. J Am Vet Med Assoc. 2015;247(8):938–44.
46. AVMA Task Force for Antimicrobial Stewardship in Companion Animal Practice. Understanding companion animal practitioners' attitudes toward antimicrobial stewardship. J Am Vet Med Assoc. 2015;247(8):883–4.
47. Redding LE, Cole SD. Pet owners' knowledge of and attitudes toward the judicious use of antimicrobials for companion animals. J Am Vet Med Assoc. 2019;254(5):626–35.
48. Smith M, King C, Davis M, Dickson A, Park J, Smith F, et al. Pet owner and vet interactions: exploring the drivers of AMR. Antimicrob Resist Infect Control. 2018;7:46.
49. Stein MR, Evason MD, Stull JW, McClure JT, Weese JS. Knowledge, attitudes and influencers of North American dog-owners surrounding antimicrobials and antimicrobial stewardship. J Small Anim Pract. 2021;62:442.

50. Weese JS. Investigation of antimicrobial use and the impact of antimicrobial use guidelines in a small animal veterinary teaching hospital: 1995-2004. J Am Vet Med Assoc. 2006;228(4):553–8.
51. Frey E, Jacob M. Commentary: using antibiograms to promote antimicrobial stewardship during treatment of bacterial cystitis and superficial bacterial folliculitis in companion animal practice. J Am Vet Med Assoc. 2020;257(9):900–3.
52. KuKanich K, Lubbers B, Salgado B. Amoxicillin and amoxicillin-clavulanate resistance in urinary Escherichia coli antibiograms of cats and dogs from the Midwestern United States. J Vet Intern Med. 2020;34(1):227–31.
53. Fowler H, Davis MA, Perkins A, Trufan S, Joy C, Buswell M, et al. A survey of veterinary antimicrobial prescribing practices, Washington state 2015. Vet Rec. 2016;179(25):651.
54. CLSI. Performance standards for antimicrobial disk and dilution susceptibility tests for bacteria isolated from animals. CLSI supplement VET08. Wayne: Clinical and Laboratory Standards Institute; 2018.
55. CLSI. Understanding susceptibility test data as a component of antimicrobial stewardship in veterinary settings. Wayne: Clinical and Laboratory Standards Institute; 2019.
56. Frey E, Costin M, Granick J, Kornya M, Weese JS. 2022 AAFP/AAHA Antimicrobial Stewardship Guidelines. J Am Anim Hosp Assoc. 2022;58(4):1–5.
57. Morley PS, Apley MD, Besser TE, Burney DP, Fedorka-Cray PJ, Papich MG, et al. Antimicrobial drug use in veterinary medicine. J Vet Intern Med. 2005;19(4):617–29.
58. Weese JS, Giguere S, Guardabassi L, Morley PS, Papich M, Ricciuto DR, et al. ACVIM consensus statement on therapeutic antimicrobial use in animals and antimicrobial resistance. J Vet Intern Med. 2015;29(2):487–98.
59. Infection Prevention and Control Best Practice Guidelines for Small Animal Veterinary Clinics. Canadian Committee on Antibiotic Resistance; 2008 [updated Aug 2008; cited 2020 Feb 6]. Available from: https://www.wormsandgermsblog.com/files/2008/04/CCAR-Guidelines-Final2.pdf.
60. Handbook of Antibiotic Stewardship in Companion Animal Veterinary Settings. University of Minnesota; 2020. Available from: https://arsi.umn.edu/handbook-antibiotic-stewardship-companion-animal-veterinary-settings.
61. Joint AVMA-FVE-CVMA statement on responsible and judicious use of antimicrobials. American Veterinary Medical Association. Available from: https://www.avma.org/resources-tools/avma-policies/joint-avma-fve-cvma-statement-responsible-and-judicious-use-antimicrobials.
62. Weese JS, Blondeau J, Boothe D, Guardabassi LG, Gumley N, Papich M, et al. International Society for Companion Animal Infectious Diseases (ISCAID) guidelines for the diagnosis and management of bacterial urinary tract infections in dogs and cats. Vet J. 2019;247:8–25.
63. Weese JS, Blondeau JM, Boothe D, Breitschwerdt EB, Guardabassi L, Hillier A, et al. Antimicrobial use guidelines for treatment of urinary tract disease in dogs and cats: Antimicrobial Guidelines Working Group of the International Society for Companion Animal Infectious Diseases. Vet Med Int. 2011;31:263768.
64. Clare S, Hartmann FA, Jooss M, Bachar E, Wong YY, Trepanier LA, et al. Short- and long-term cure rates of short-duration trimethoprim-sulfamethoxazole treatment in female dogs with uncomplicated bacterial cystitis. J Vet Intern Med. 2014;28(3):818–26.
65. Jessen LR, Sorensen TM, Bjornvad CR, Nielsen SS, Guardabassi L. Effect of antibiotic treatment in canine and feline urinary tract infections: a systematic review. Vet J. 2015;203(3):270–7.
66. Westropp JL, Sykes JE, Irom S, Daniels JB, Smith A, Keil D, et al. Evaluation of the efficacy and safety of high dose short duration enrofloxacin treatment regimen for uncomplicated urinary tract infections in dogs. J Vet Intern Med. 2012;26(3):506–12.
67. Hillier A, Lloyd DH, Weese JS, Blondeau JM, Boothe D, Breitschwerdt E, et al. Guidelines for the diagnosis and antimicrobial therapy of canine superficial bacterial folliculitis (Antimicrobial Guidelines Working Group of the International Society for Companion Animal Infectious Diseases). Vet Dermatol. 2014;25(3):163–e43.

68. Lappin MR, Blondeau J, Boothe D, Breitschwerdt EB, Guardabassi L, Lloyd DH, et al. Antimicrobial use guidelines for treatment of respiratory tract disease in dogs and cats: Antimicrobial Guidelines Working Group of the International Society for Companion Animal Infectious Diseases. J Vet Intern Med. 2017;31(2):279–94.
69. Dear JD. Bacterial pneumonia in dogs and cats: an update. Vet Clin North Am Small Anim Pract. 2020;50(2):447–65.
70. Wayne A, Davis M, Sinnott VB, Bracker K. Outcomes in dogs with uncomplicated, presumptive bacterial pneumonia treated with short or long course antibiotics. Can Vet J. 2017;58(6):610–3.
71. Fernandes Rodrigues N, Giraud L, Bolen G, Fastres A, Clercx C, Gommeren K, et al. Antimicrobial discontinuation in dogs with acute aspiration pneumonia based on clinical improvement and normalization of C-reactive protein concentration. J Vet Intern Med. 2022;36(3):1082–8.
72. Vientos-Plotts AI, Masseau I, Reinero CR. Comparison of short- versus long-course antimicrobial therapy of uncomplicated bacterial pneumonia in dogs: a double-blinded, placebo-controlled pilot study. Animals. 2021;11(11):3096.
73. Littman MP, Gerber B, Goldstein RE, Labato MA, Lappin MR, Moore GE. ACVIM consensus update on Lyme borreliosis in dogs and cats. J Vet Intern Med. 2018;32(3):887–903.
74. Sykes JE, Hartmann K, Lunn KF, Moore GE, Stoddard RA, Goldstein RE. 2010 ACVIM small animal consensus statement on leptospirosis: diagnosis, epidemiology, treatment, and prevention. J Vet Intern Med. 2011;25(1):1–13.
75. Morris DO, Loeffler A, Davis MF, Guardabassi L, Weese JS. Recommendations for approaches to meticillin-resistant staphylococcal infections of small animals: diagnosis, therapeutic considerations and preventative measures. Clinical consensus guidelines of the World Association for Veterinary Dermatology. Vet Dermatol. 2017;28(3):304-e69.
76. Neer TM, Breitschwerdt EB, Greene RT, Lappin MR. Consensus statement on ehrlichial disease of small animals from the infectious disease study group of the ACVIM. American College of Veterinary Internal Medicine. J Vet Intern Med. 2002;16(3):309–15.
77. Hopman NEM, Portengen L, Heederik DJJ, Wagenaar JA, Van Geijlswijk IM, Broens EM. Time trends, seasonal differences and determinants of systemic antimicrobial use in companion animal clinics (2012-2015). Vet Microbiol. 2019;235:289–94.
78. Redding LE, Grunwald H, Melofchik C, Meily P, Henry A, Stefanovski D. Comparison of animal daily doses and days of therapy for antimicrobials in species of veterinary importance. Prev Vet Med. 2020;176:104942.

Antimicrobial Stewardship in Immunocompromised Hosts

7

Tine Vindenes, Hrvoje Melinscak, Kristin Linder, and Majd Alsoubani

The importance of antimicrobial stewardship (AMS) in hospitals is well established, and yet there remain specialized care settings in which implementation is particularly difficult. The care of immunocompromised hosts is one of those settings. Highly vulnerable and leaving little room for error, it can be difficult for prescribers to forgo, de-escalate, or discontinue antimicrobials when such an approach would be indicated for other patients. Compounding these challenges is an absence of controlled clinical trials and national or international guidelines. At the same time, multidrug-resistant (MDR) infections are on the rise and can be devastating in terms of outcomes especially in the immunocompromised, are costly to the healthcare system, and are a threat to the advancement of medicine and overall public health. Immunosuppressed patients, whether they be oncology patients, people living with HIV/AIDS (PLWHA), or transplant recipients, have unique characteristics and are generally at higher risk of infections. The principles of rational and thoughtful AMS may be adapted to good effect in such populations. In this chapter, we describe the landscape surrounding and approach to AMS in three patient populations: oncology, HIV/AIDS, and transplantation (both solid organ and hematopoietic stem cell).

T. Vindenes (✉) · M. Alsoubani
Division of Geographic Medicine and Infectious Diseases, Tufts Medical Center, Boston, MA, USA
e-mail: tvindenes@tuftsmedicalcenter.org; malsoubani@tuftsmedicalcenter.org

H. Melinscak
Division of Hematology and Oncology, Tufts Medical Center, Boston, MA, USA
e-mail: hmelinscak@tuftsmedicalcenter.org

K. Linder
Division of Pharmacy, Hartford HealthCare, Hartford, CT, USA
e-mail: Kristin.linder@hhchealth.org

S. Doron, M. Campion (eds.), *Antimicrobial Stewardship in Non-Traditional Settings*, https://doi.org/10.1007/978-3-031-21710-4_7

Antimicrobial Stewardship in Oncology

Febrile neutropenia and infection are among the most frequent and serious complications of chemotherapy. Approximately 8 per 1000 patients receiving chemotherapy develop febrile neutropenia, leading to greater than 60,000 hospitalizations and 4000 deaths per year in the USA [1, 2]. Patients with hematologic malignancies are at higher risk of developing febrile neutropenia, occurring at a rate of approximately 43 per 1000 cases [3]. The mortality rate among patients hospitalized for febrile neutropenia has been estimated to be between 2% and 21% [2]. The cost of caring for this population is significant. A national surveillance study in 2012 found the total cost of cancer-related neutropenia hospitalization to be $2.3 billion for adults and $439 million for children among 91,560 and 16,859 admissions for adults and children, respectively [2]. This equates to a mean cost of $24,770 per hospitalization for adults and $26,000 per stay for children [2]. Febrile neutropenia is defined as a single oral temperature greater than 38.3 °C or two temperature readings greater than 38.0 °C for more than 1 h and an absolute neutrophil count (ANC) of $<0.5 \times 10^4$/l, or expected to fall below $<0.5 \times 10^4$/L. Due to the complexity of this patient population, clinically documented infection is identified only 20–30% of the time [4]. Bacteremia occurs in 10–25% of all neutropenic patients and is associated with prolonged or profound neutropenia (ANC < 100 neutrophils/mm) [3, 4]. Less commonly, neutropenic patients can also develop infections caused by fungi and viruses. However, the focus of this section will be on bacterial pathogens. Many reports have documented that infections caused by drug-resistant Gram-positive and Gram-negative bacteria are increasing in febrile neutropenic patients, highlighting the need for AMS [5–8]. This increase is likely a result of the high frequency of infection in this patient population and corresponding antimicrobial treatment and prophylaxis. Implementation of AMS initiatives and infection control measures are the cornerstone for controlling the development and spread of antimicrobial resistant pathogens in this patient population.

Antimicrobial Resistance in Oncology Patients

The development of antimicrobial resistance has become a significant global problem, and cancer patients are among those most significantly affected. Although there is substantial geographic variability in rates of antibiotic resistance among cancer patients, the frequency of bloodstream infections caused by MDR organisms has been reported to be as high as 35% in some institutions with mortality rates of 35–80% [9, 10]. Epidemiological studies indicate rising resistance rates to common empiric beta-lactam antibiotics [11, 12]. Six organisms have been identified as concerning due to the ability to become resistant or "escape" the treatment of many antimicrobials: *Enterococcus faecium*, *Staphylococcus aureus*, *Klebsiella pneumoniae*, *Acinetobacter baumanii*, *Pseudomonas aeruginosa*, and *Enterobacter* spp. (the "ESKAPE" pathogens) [13]. These organisms cause the majority of nosocomial infections in oncologic patients. A more recent acronym, "ESKAPEES," adds

Escherichia coli (E.coli) and *Stenotrophomonas maltophilia*, which are also clinically relevant in the neutropenia patient population [14].

Risk factors for antimicrobial resistance in the immunocompromised and cancer patient populations have been found to be similar to those in immunocompetent hosts. Specific risk factors can determine the likelihood of a pathogen. Generally, risk factors associated with infection with a resistant organism in patients with malignancy include history of broad-spectrum antimicrobial use (specifically third-generation cephalosporins, carbapenems, fluoroquinolone prophylaxis, and vancomycin), intensive care unit (ICU) stay, invasive procedures, prolonged or recurrent hospitalization, use of urinary or indwelling catheters, older age, colonization with resistant organisms, and presence of comorbidities [15–27].

Gram Negative Bacterial Resistance in Oncology

The etiology of bloodstream infection (BSI) in cancer patients in the last decade has undergone important shifts in the spectrum of bloodstream isolates. Gram-negative organisms are becoming more prevalent and in some areas overtaking Gram-positive bacteria in frequency as the cause of bacteremia [11, 12, 28, 29]. The incidence of resistance among Gram-negative organisms causing bacteremia has also risen [11, 28, 29]. The three most frequently isolated Gram-negative bacilli (GNB) in neutropenic patients are *E. coli*, *P. aeruginosa*, and *Klebsiella* spp. These organisms account for 45–60% of documented GNB infections [28, 30, 31]. Other less common *Enterobacteriaceae* implicated in neutropenic patients include *Citrobacter* spp., *Enterobacter* spp., *Proteus* spp., and *Serratia* spp. The incidence of these pathogens is likely to be institution specific.

Antimicrobial resistance among *Enterobacteriaceae* in cancer and hematopoietic stem cell transplant (HSCT) recipients has been described mainly due to the production of beta-lactamases including Ambler class C (AmpC) and the extended-spectrum beta-lactamases (ESBL). Coexisting resistance genes conferring resistance to multiple classes of antimicrobials are commonly found on the same plasmid. The incidence of *Enterobacteriaceae* carrying ESBL is quite variable, dependent on location and patient population studied. ESBL is the most commonly seen resistance mechanisms in oncology patients, primarily in *E. coli* and *Klebsiella* spp. [12] Carbapenem-resistant Enterobacteriaceae (CRE) are also becoming an emergent threat, but the incidence is not yet well described in the oncology patient population. Carbapenem resistance results from production of a carbapenem-hydrolyzing enzyme or changes in outer membrane porins combined with ESBL and AmpC overproduction. A more recent study in patients with hematologic malignancies highlights that an absence of previous carbapenem use does not preclude the development of infection due to CRE, highlighting the ability of organisms to acquire resistance through clonal spread [32].

Multidrug-resistant (MDR) pseudomonas is often a result of multiple different mechanisms, often involving Amp-C cephalosporinase production and changes in outer membrane porins. Infection due to MDR pseudomonas in patients with cancer has been associated with inadequate empiric antimicrobial therapy and with increased morbidity and mortality [30].

Two other less common but clinically relevant Gram-negative organisms reported in this patient population are *S. maltophilia* and *A. baumanii*. Increasing rates of both organisms have been reported in immunocompromised patients [15, 33–35]. *A. baumanii* in particular has been associated with nosocomial outbreaks in intensive care units. Treatment of both organisms tends to be challenging as *S. maltophilia* is inherently resistant to most antibiotics and *A. baumanii* can quickly develop resistance through multiple mechanisms.

Gram-Positive Bacterial Resistance in Oncology

Infections due to staphylococci, streptococci, and enterococci remain a clinically significant problem in the oncology patient population. The most common Gram-positive pathogens isolated from neutropenic patients include coagulase-negative staphylococci (CoNS), *S. aureus*, viridans group strep (VGS), and Enterococcus species.

Methicillin resistance among *S. aureus* is due to the mecA gene encoding for a low-affinity penicillin-binding protein, PBP2a. Greater than 90% of CoNS and up to 80% of *S. aureus* are methicillin resistant [14]. Resistance among *S. aureus* varies widely based on geography and institution. Methicillin-resistant *S. aureus* has increased in the last two decades; however, recent surveillance data indicate that rates may actually be decreasing or leveling out [36, 37]. Concerning increases in minimum inhibitory concentrations (MICs) and treatment failures with vancomycin have been reported. However, there is conflicting data to support this phenomenon. Current evidence suggests that this MIC creep phenomenon is more of an institution-specific finding [38].

Antibiotic resistance among VGS varies dependent on institution and geographic location. Recent studies report a 4–14% incidence of highly resistant penicillin VGS isolates in the neutropenic patient population [39, 40]. Within this group of organisms, *Streptococcus mitis* is most consistently penicillin resistant, with rates reported to be 50–86% in pediatrics [41].

Enterococci are a significant cause of bacteremia in the oncology patient population, specifically patients undergoing stem cell transplantation or those with hematologic malignancy. It has been reported that up to 35.5% of nosocomial enterococcal infections are vancomycin resistant, with *Enterococcus faecium* accounting for the majority of infection [42]. Vancomycin resistance is mediated by two types of gene clusters (*van*A and *van*B), which confer resistance by altering the target site for vancomycin. Increasing rates of VRE have led to increased use of alternative agents, including daptomycin and linezolid. Daptomycin failure and resistance development have been reported for VRE in the treatment of deep-seated enterococcal infections. However, failure was likely due to using dosing schemes approved for MRSA infection, which has a lower MIC breakpoint [38]. Linezolid remains reliably active against VRE, but adverse event profile (thrombocytopenia) and a lack of evidence from clinical trials of its effectiveness against severe infections in this population limit its use in the immunocompromised.

Preventing Resistance

Antimicrobial stewardship and infection control measures are the cornerstone for controlling the development and spread of antimicrobial resistance. As new antimicrobial agents have become available with activity for resistant organisms (ceftolozane/tazobactam, ceftazidime/avibactam, meropenem/vaborbactam, cefiderocol), institutions should outline clear restrictions for use and attempt to identify patients at increased risk for resistance based on previous culture history and antibiotic use. Immunocompromised patients are often excluded from clinical drug trials, leaving it to infectious disease physicians and subject matter experts to determine how to use these new drugs in those populations.

Antimicrobial Stewardship Principles in Oncology

Antimicrobial stewardship has emerged as a vital strategy to optimize antimicrobial use; it aims to limit the emergence of resistant organisms, minimize medication toxicity, minimize development of *Clostridioides difficile* infection, and contain cost [43]. These lofty ideals are achieved, in no small part, by giving due consideration to the antimicrobial regimen selected, in addition to appropriate dosing, duration, route of delivery, and de-escalating or escalating antibiotics where appropriate. In oncology, polypharmacy is not infrequent; thus, it is essential to assess for drug–drug interactions, and it is also important to review the ongoing need for medications prescribed (antibiotics included). Revisiting the utility in continuing antibiotics frequently is a fundamental principle that can be applied to practice readily. There is a re-energized effort to explore the rational use of antimicrobials while examining the merits of short course therapy for its efficacy and lesser risk of adverse effects relative to longer duration treatment [44]. It remains unclear under which circumstances this might be applied to oncology practice.

Indeed, there are additional means by which AMS might achieve its ends to help diminish infections and reduce antibiotic use. One simple measure involves scrutinizing cultures closely for potential contaminants, or mere colonization, vs. true infection. Obtaining appropriate cultures based upon the signs and symptoms is preferred to "pan-culturing." For instance, identification of more than two organisms in a urine culture suggests contamination. Further, a positive urine culture without a fever or other signs or symptoms of a urinary tract infection often depicts colonization. Naturally, a rational appraisal of the clinical picture in context is important for those individuals with stents, nephrostomies, and conduits, who require more thoughtful consideration where the aforementioned may not hold true [45].

Other gestures as simple as liaising with the local microbiology laboratory, surveying local speciation and sensitivity patterns, or referring to hospital-specific antibiograms provide valuable insight and direct therapy more effectively. An antibiogram is an aggregate summary of the antibiotic susceptibility data for bacterial isolates from a microbiology laboratory over a defined period (typically 1 year). In 2017, the Joint Commission mandated AMS programs for all hospitals [46]. In Massachusetts, hospital-specific antibiogram reporting has been required of all acute care hospitals since 2013 [47]. Such mandated reporting has expanded our understanding of statewide microflora and their resistance patterns. Hospital-specific antibiograms are meant to guide all clinicians in terms of empiric antibiotic therapy selection, and once culture data are finalized subsequent review allows tailoring of regimens thus. Further study is required to assess whether hematology-oncology ward-specific antibiograms might improve care and outcomes. More intriguing still is the concept of patient-specific antibiograms which would account for individual risk factors and adapt an algorithmic approach to better predict future susceptibility patterns in infections experienced by a given patient [48].

Another area of focus concerning AMS involves antibiotic transitions. Appropriate intravenous to oral swaps or initiating therapy with oral antibiotics of high bioavailability may reduce cost, reduce length of hospital stay, and prevent the risks incurred by placing venous access (infection, bleeding, or thrombosis) [43]. Extending this approach to the oncology patient is reasonable with certain caveats applied. Nausea, vomiting, and diarrhea due to chemotherapy or alterations in gut flora may impact absorption of certain antibiotics. A subset of cancer patients may suffer gastrointestinal obstruction precluding safe intake of oral medication. It must be said that oncology patients, as a whole, may already be equipped with vascular access, and enjoy frequent outpatient follow-up; however, cost savings may yet be realized where narrowing of spectrum and transition to oral antimicrobials is carried out.

Antibiotic allergies are present in all patient populations. A careful investigation of medication allergy might include the input of an allergy consultant and early skin testing batteries. Studies have demonstrated that immunocompromised patients resigned to penicillin or beta-lactam alternatives suffer adverse outcomes which contribute increased cost to their hospital stay [49]. Penicillin testing in the immunocompromised host has been shown to be safe and effective when assessing for penicillin or beta-lactam allergies and is therefore a key element of AMS [50].

Febrile Neutropenia and Antimicrobial Stewardship

Febrile neutropenia poses an imminent threat to the oncology patient and is considered an oncologic emergency. Neutrophils constitute an essential host defense against bacterial and fungal infections; moreover, broad-spectrum antimicrobial use is the rule where fever and neutropenia intersect, yet questions of choice of agent or agents and duration of therapy remain the source of ongoing consternation. Current standards of care for empiric antimicrobial therapy of febrile neutropenia depend upon an initial risk assessment (as per the 2022 National Comprehensive Cancer

Network, NCCN, guidelines) [51]. Those considered at low risk of developing complications (as assessed by the Multinational Association for Supportive Care in Cancer risk index score, elements of which include an anticipated short duration of severe neutropenia, absence of organ dysfunction, and good performance status) presenting with fever and neutropenia may be well served by an outpatient care plan consisting of antibiotics and careful follow-up. A critical appraisal of a patient's laboratory studies, social support structure, consent for home care, and eligibility for oral antimicrobial therapy (has no nausea or vomiting, may tolerate oral medications, and no prior fluoroquinolone therapy) is imperative [51]. In this low-risk group treatment options include intravenous antibiotics administered at home, a long-acting intravenous antibiotic administered either at home or in office along with an oral antibiotic, or oral antibiotic alone (moxifloxacin, levofloxacin, ciprofloxacin, or amoxicillin/clavulanate, with clindamycin as an alternative if allergic to penicillin and cephalosporin). If the patient is considered at high risk for serious complications, then inpatient hospital care is recommended. The recommended empiric antibiotic should be based on infection risk assessment and presumptive site of infection, local antibiogram data, organ dysfunction, prior antibiotic treatment, risk of multidrug-resistant organisms, antibiotic allergies, and colonization with or prior infection with methicillin-resistant *Staphylococcus aureus* (MRSA) [51]. While anti-pseudomonal Gram-negative therapy is essential for inpatient treatment of fever and neutropenia, there are only a few sources where anti-MRSA therapy is also recommended in the empiric regimen according to the NCCN guidelines [51]: sinus/nasal infection, periorbital cellulitis, bacterial pneumonia, vascular access devices, and central nervous system infections [51].

It is oft a simple matter to initiate antimicrobial therapy; however, it is more daunting to produce quality evidence to support a specific duration of treatment especially where there is a lack of clear clinical findings and microbiologic data. NCCN guidelines relating to the duration of therapy in the absence of clinical and microbiologic data following resolution of fever are based on a neutrophil count threshold of 500 cells/mcL. For instance, an absolute neutrophil count (ANC) of less than 500 cells/mcL in the face of fever resolution might have one either de-escalate to a prophylactic regimen, or continue current therapy until count recovery, or discontinue therapy altogether [51]. Such niche guidelines often equivocate for the literature has been slow to advance; however, a small cluster of recent studies regarding patients with malignant hematological diseases have challenged long-standing canon as it pertains to neutropenic fever. A small, prospective, European observational study ANTIBIOSTOP [52] assessed the feasibility and safety of short-term antibiotic treatment in afebrile or febrile patients without a clear source irrespective of neutrophil count. Antibiotics were stopped after a 48-h fever free period [$n = 45$] in the first phase of the study; in the second phase of the study, antibiotic therapy was discontinued in no later than 5 days for all fevers of unknown origin irrespective of temperature or leucocyte counts. There was no significant difference between phases of the study in terms of recurrent fevers, intensive care unit admissions, infection relapse 48 h after stopping antibiotics, or in-hospital mortality. The study concluded that early discontinuation of empiric antibiotics in patients with fever of unclear origin is safe for afebrile neutropenic patients, but that a larger randomized study is needed to confirm the safety of stopping antibiotics after 5 days irrespective of temperature curve and neutrophil count. Another small retrospective study assessed recurrent fever in adult hematopoietic stem cell transplant patients with neutropenia after early de-escalation of antimicrobial therapy [53]. Patients in the intervention cohort enjoyed de-escalation from broad-spectrum antibiotics and were transitioned to prophylactic oral antibiotics after receiving 5 days of therapy and remaining afebrile for 48 h. Compared with patients receiving broad-spectrum antibiotics until neutrophil recovery, the study found no increase in recurrent fever in the early de-escalation group compared to the no de-escalation group, respectively (15% vs. 19%, $p = 0.26$). There were no statistical differences between cohorts in the need for antibiotic escalation, intensive care admissions, bacteremia, or in-hospital mortality. There was an overall trend toward reduction in cost as well as less Gram-negative antibiotic utilization in the early de-escalation group [53]. Similarly, an open-label randomized study, prospective in nature (the so-called "How Long Study"), randomized adults with hematologic malignancies suffering neutropenic fever to either an early antimicrobial discontinuation arm (requiring 72 h spent fever free and with clinical recovery) or a conventional strategy of antimicrobial discontinuation when the neutrophil count rose above 500 cells/mcL. The study compared empiric antimicrobial therapy free days between arms [54]. High-risk individuals with febrile neutropenia and without etiology for their fever were treated according to local protocols and international guidelines. Those in the early de-escalation arm enjoyed a greater number of empiric antibiotic free days; moreover, severe (non-bacterial) adverse events were fewer. Mortality and days with fever were not significantly different, and no oral prophylactic antibiotics were employed in this particular study [54]. These small studies add to a growing body of evidence suggesting that perhaps less is more in the immunocompromised host and that earlier de-escalation of empiric broad-spectrum antibiotics is safe in cases of febrile neutropenia without a known source. It should be again mentioned that 20–30% of patients with febrile neutropenia have no clinically documented infection; this raises the intriguing possibility that many of these patients may not have an infection, but rather suffer from non-infectious fevers for various reasons.

Antimicrobial Prophylaxis in Oncology

Modern hematology-oncology practice has come to regard antimicrobials as staple in the context of comprehensive cancer care. The addition of prophylactic antimicrobials facilitates the use of particularly stringent chemotherapy regimens (largely with curative intent) that often lead to a significant and prolonged neutropenia. Without such prophylactic therapy, febrile neutropenia (FN) would pose a formidable obstacle to meaningful cancer treatment particularly in the domain of hematologic malignancy. Solid tumors and their treatment, on the other hand, are not so much plagued by enduring immunosuppression, but are often accompanied by unique circumstances, such as obstructive phenomena (pulmonary and hepatobiliary tracts being pre-eminent in this regard), or surgical maneuvers that might be complicated by infection, as well as indwelling appliances that offer a handy portal of entry to any willing invader. In addition, chemotherapy can lead to mucositis and disturb the microbiome, and immune-modulating viruses such as cytomegalovirus may occasionally alter the risk of infection [55].

The 2018 joint American Society of Clinical Oncology (ASCO) and Infectious Disease Society of America (IDSA) publication guideline update regarding antimicrobial prophylaxis for adult patients with cancer-related immunosuppression emphasized the risk assessment of febrile neutropenia first and foremost [55]. Patient characteristics, underlying malignancy, and treatment of said malignancy all play a part in risk assessment. Host factors consisted of age, performance status, nutritional status, prior FN episode, and comorbidities. The underlying malignancy category included features of cancer diagnosis (the reported FN rates varying from 1 to 95% between the various cancer diagnoses), cancer stage, remission status, and cancer treatment response. The treatment of malignancy category factored in cytotoxic regimen employed, dose intensity, degree and duration of gastrointestinal and oral mucositis, and degree and duration of cytopenias [55]. The guideline recommends antibiotic prophylaxis consisting of a fluoroquinolone during periods of neutropenia for patients who are deemed high risk for FN or profound, protracted neutropenia. For example, a high-risk patient would be considered one who suffers acute leukemia and is undergoing treatment with a myeloablative conditioning regimen. On the other hand, since treatment of solid tumors generally does not result in severe and prolonged periods of neutropenia, the risk of FN is lower and thus fluoroquinolone prophylaxis is not recommended. Despite the guideline's attempt to forward a detailed risk assessment tool, it remains difficult to assess risk related to FN in a given patient undergoing cytotoxic chemotherapy. There is sufficient ambiguity to muddy the decision of whether a patient's risk is high enough to warrant prophylactic antimicrobial treatment. How high a risk must be to warrant antimicrobial prophylaxis (that outweighs the risk of antimicrobial use) is up to the consumer of such guidelines, and there is no definitive guide as to how to weigh the given risk factors. In short, this is not a scoring calculator, but rather an instrument to complement sound clinical judgment.

A few concerns regarding the outright need for and use of antimicrobial prophylaxis have been raised in the literature [56]. A small, but relevant, German study included 63 patients undergoing allogeneic stem cell transplantation. No antimicrobial prophylaxis was administered, and there was no resultant increase in mortality or bloodstream infections compared to existing published data for patients prescribed fluoroquinolone prophylaxis [57]. ASCO/IDSA guidelines for prophylaxis in cancer-related immunosuppression reference the rising rates of fluoroquinolone resistance; moreover, fluoroquinolone resistance in *Enterobacteriaceae* could well be as high as 30% in certain communities [55]. It is therefore reasonable to question at what level of *Enterobacteriaceae* resistance does prophylaxis become useless and offer only the potential for harm? [55] Notwithstanding, Kern and company noted that resistance rates [alone] were a poor indicator of the potential clinical benefit associated with use of prophylaxis [58]. Might the use of prophylaxis be better determined by local resistance rates and local antibiograms in those patients at high risk of FN? [59] Multiple studies have shown significant adverse impact upon the microbiota when using broad-spectrum antibiotics, fluoroquinolones included, which may well lead to adverse outcomes in the cancer patient. These changes in microbiota may not only result in *Clostridioides difficile* infection (see below), but can potentially alter the gut flora such that it harbors resistant Gram-negative microbes [60]. Furthermore, these alterations may select for Gram-positive bacteria, such as *Enterococci*, which later foster invasive infections [60–64]. An often underappreciated consequence of alterations in the gut microbiome is the loss of microbial diversity in those suffering graft-versus-host disease and the resultant impact on outcomes [65]. One final consideration is that the etiology of bloodstream infections (BSIs) in patients with cancer has shifted from that of primarily Gram-negative organisms toward Gram-positive organisms, and back again (lately favoring Gram-negative organisms once again). The difference this time, however, is the increased resistance of those Gram-negative organisms. The multiple reasons for changes in the etiology of BSIs over time are beyond the scope of this discussion, but suffice it to say, the use of fluoroquinolone prophylaxis has likely played a significant role in these patterns.

Clostridioides Difficile **in Oncology Patients**

Clostridioides difficile infection (CDI) is associated with higher incidence, mortality rates, and length of stay in patients with cancer compared to those without. Recent studies found that the incidence in the general hospitalized patient population is 1–2% compared with 7–14% in individuals with a cancer diagnosis and 7% in individuals undergoing chemotherapy [66, 67]. Similar findings have been observed in the pediatric cancer population [68]. Patients with a higher degree of immunosuppression due to hematologic malignancy or hematopoietic stem cell transplantation (HSCT) appear to have a greater risk. In a recent report, the incidence of CDI among HSCT recipients was nine times greater than the general population and 1.4 times greater than the oncology patient population with no history of HSCT [69]. Additionally, a national study assessing CDI incidence in oncology patients using the US National Hospital Discharge Surveys (NHDS) from 2001 to 2010 found that the incidence of CDI was higher among hematologic cancers, with 17.3 infections per 1000 discharges compared to 6.8 infections per 1000 discharges in patients with solid tumors [70]. The same study found higher mortality and longer median length of stay for cancer patients with CDI compared to without CDI. Elevated mortality rates (9.1–19.7%) and longer lengths of stays among cancer patients with CDI compared to patients without CDI (both cancer and non-cancer) have been consistently reported in the literature [70–74]. Given the particularly high risk of CDI and poor outcomes among cancer patients, appropriate diagnosis, treatment, and prevention of further recurrences are of utmost importance.

Clostridioides difficile is an anaerobic Gram-positive and spore-forming bacterium that typically exists as a nonpathogenic colonizer in 7–18% of the adult population [75]. Disease associated with *Clostridioides difficile* secondary to antibiotic use was first reported in 1978 [76]. Shortly after in 1981, the first case of chemotherapy-associated CDI was reported in a patient receiving cisplatin, bleomycin, and vincristine for testicular embryonal cell cancer without previous antibiotic exposure [77]. Since this time, a growing amount of evidence has linked antibiotic use, chemotherapy, and other risk factors with the development of *C. difficile* infection. The organism's pathogenesis and pathophysiology lend to a variety of risk factors in the oncology patient population. *Clostridioides difficile* causes disease when three major conditions in the host are met: (1) Disruption of the normal diversity of the gut microbiome, (2) colonization of the host with a pathogenic strain capable of producing toxins from an exogenous source (either preceding or following microbiome disruption), and (3) transformation of *C. difficile* bacteria into a vegetative state, which activates toxin gene transcription and leads to production of toxins A (enterotoxin) and B (cytotoxin) [78]. Toxins A and B then bind to epithelial cells in the intestines, where they are internalized and mediate reactions leading to cell death.

The most important risk factors for the development of CDI include any factor that can alter the gut microbiome and increase exposure to pathogenic *Clostridioides difficile* strains. Exposure to antibiotics has been considered one of the main risk factors for development of CDI. Oncology patients generally are at higher risk for infection and therefore are more likely to receive therapeutic or prophylactic antimicrobials compared to patients without underlying cancer [79]. Additionally, complications of malignancy and chemotherapy lead to increased antimicrobial use. Febrile neutropenia is often treated with broad-spectrum antimicrobials until recovery of neutrophils, which can take anywhere from days to weeks depending on the type of cancer and chemotherapy. Patients at high risk for infectious complications, including patients with hematological malignancies and expected duration of neutropenia greater than 7 days, often receive prophylactic antibiotics and antivirals at the start of chemotherapy or once neutropenia occurs.

Chemotherapy has been identified as a possible independent risk factor for CDI, even in the absence of antimicrobial use [80]. Cytotoxic chemotherapy has antibiotic activity in the gastrointestinal tract and can lead to changes in the gut microbiome. In addition, chemotherapy can lead to immunosuppression and direct mucosal damage to epithelial cells in the gastrointestinal tract, which is likely to promote CDI [80–83]. Pharmacologic classes of chemotherapeutics that have been associated with development of CDI in the literature include antimetabolites, DNA topoisomerase inhibitors, platinum-based agents, taxanes, and alkylating agents [80, 81, 84–86].

Other risk factors for CDI identified in the oncology patient population include frequent and prolonged hospitalization, advanced age, gastric acid suppressing medications, manipulation of the gastrointestinal tract (i.e., abdominal surgery, enteral feeding), and radiation therapy [80, 87–91].

Diagnosis of CDI in the oncology patient population can be difficult to distinguish from diarrhea secondary to chemotherapy or diarrheal illness caused by other intestinal pathogens. Therefore, accurate diagnosis techniques are essential for identifying patients with disease to ensure appropriate treatment and infection prevention measures, which has been associated with reduced mortality and overall healthcare costs [77, 92]. Specific to the oncology patient population, the initial signs and symptoms of CDI when associated with chemotherapy typically occur 1 week after receiving treatment [93]. Timing of CDI onset with respect to concomitant antimicrobial therapy or other risk factors has not been well described. However, the risk for infection in patients with malignancy generally increases with the duration and degree of neutropenia. According to the Society for Healthcare Epidemiology of America (SHEA) and Infectious Diseases Society of America (IDSA) Guidelines, most recently updated in 2018, patients experiencing new-onset diarrhea with greater than or equal to three unformed stools in the previous 24 h should be tested for *C. difficile* infection [94]. Oncology patients who meet these criteria should be tested regardless of antimicrobial use or chemotherapy history.

Multiple diagnostic tests exist to aid in the diagnosis of CDI. However, a significant limitation of current testing methods is the inability to accurately differentiate active disease from colonization. As such, oncology patients are at an increased risk for overtreatment, which may lead to further intestinal dysbiosis and may increase the risk for colonization with vancomycin-resistant enterococcus [95]. Current IDSA guidelines emphasize the importance of testing appropriate stool samples (i.e., only in symptomatic patients not receiving laxatives). In the absence of appropriate laboratory stewardship programs for CDI, multistep algorithms are recommended. However, in the setting of appropriate stool testing, PCR alone or a multistep algorithm for testing can be used. Additionally, a GI multiplex panel may be considered to aid in the diagnosis of other gastrointestinal pathogens including norovirus, adenovirus, and rotavirus.

Treatment of CDI in oncology patients is guided by current national guidelines which direct initial therapy on the basis of disease severity. Multiple scoring systems exist for classifying severity of illness for CDI [94, 96]. According to the IDSA/SHEA treatment guidelines, severe disease is defined by a white blood cell (WBC) count greater than 15,000 cells/μL or a serum creatinine level greater than 1.5 mg/dL. The American College of Gastroenterology (ACG) guidelines have slightly different criteria, consisting of a serum albumin less than 3 g/dL with either an elevated WBC count of 15,000 cells/μL or abdominal tenderness [94, 96].

Current IDSA guidelines recommend oral vancomycin or fidaxomicin for an initial episode of CDI for all patients, with the 2021 update to the guideline favoring fidaxomicin due to data suggesting decreased CDI recurrences following fidaxomicin treatment [97]. Treatment with metronidazole is no longer recommended first line. Even before metronidazole fell out of favor, the National Comprehensive Cancer Network (NCCN) guidelines recommended vancomycin as the preferred treatment agent and metronidazole and fidaxomicin as alternatives [51]. This recommendation was a result of a report in 2005 indicating decreased clinical cure rates with metronidazole [98]. More recent studies found similar results, supporting the use of oral vancomycin or fidaxomicin as first-line therapy in patients presenting with an initial episode of either severe or non-severe CDI [99, 100].

Notably, a subgroup analysis from the two large randomized controlled trials illustrating non-inferiority of fidaxomicin compared to vancomycin evaluated efficacy of both agents in patients with active malignancy [101]. Among patients with active malignancy the overall cure rate was found to be significantly lower (79.2% vs. 88%, $P < 0.001$) and the time to resolution of diarrhea was shown to be delayed (100 h vs. 55 h) when compared to patients without cancer in these studies. While patients with malignancy were found to have a more rapid resolution of diarrhea in the fidaxomicin group, this conclusion has not been validated in other studies. The NCCN guidelines continue to recommend vancomycin and fidaxomicin as first-line therapies.

According to national guidelines, recurrent infection should be treated with fidaxomicin (standard or extended-pulsed regimen). Fecal microbiota transplant is recommended in the setting of multiple recurrences and treatment failure of standard treatment. The recommendation for FMT in recurrent CDI is supported by multiple anecdotal reports as well as more recently, randomized controlled trials. Efficacy rates for FMT in the literature have varied from 50% to 90% [102]. However, notable studies excluded patients with neutropenia and recent chemotherapy and therefore should be interpreted cautiously. Although there are limited data on the use of FMT in the oncologic patient population, a recent multicenter case series of 80 immunocompromised patients showed FMT to be safe in this patient population [103]. Additionally, in a small report of seven HSCT undergoing FMT, Webb et al. [104] reported efficacy in six out of seven patients without severe side effects or mortality. Although oncology patients are certainly at increased risk of infection due to underlying disease and therapy, to date there have been few reports of infectious complications secondary to FMT and no identified reports in the oncologic patient population. Based on these findings, many experts support the use of FMT in immunocompromised patients in the setting of treatment failure and frequent recurrences. Bezlotoxumab has also been evaluated for recurrent CDI in the oncology patient population. Bezlotoxumab is a human monoclonal antibody against *Clostridioides difficile* toxin B. It currently has an indication to prevent CDI recurrence in adults at high risk for recurrent infection. The 2021 update to the IDSA treatment guidelines for CDI recommends use of bezlotoxumab for patients with a recurrent CDI episode within the last 6 months [97]. A recent post hoc analysis of the phase III trial data specifically evaluated outcomes associated with bezlotoxumab administration in patients with specific risk factors for recurrence, including age ≥ 65, history of CDI, immunocompromised status, severe CDI, and ribotype 027/078/244 [105]. The analysis found that patients with three risk factors for recurrence or greater benefited the most from receiving bezlotoxumab therapy, illustrating an absolute risk reduction of 24.8%. However, patients with one or two risk factors also benefited from therapy. Since patients with malignancy may have several risk factors for recurrence, bezlotoxumab may play an important role for prevention in this patient population.

Prophylaxis of *C. difficile* in Oncology Patients

Probiotics as primary prevention for CDI have been evaluated in several randomized controlled trials and meta-analyses with varying results. Current national guidelines provide no guidance on the use of probiotics for primary prevention due to insufficient data and limitations of the current literature [94]. The utility of probiotics in the oncology patient population has not been well studied. However, bloodstream infections caused by common probiotic organisms have been described in the literature in both the immunocompromised and immunocompetent patient populations [106–109]. Given the questionable benefits and theoretical risks of these products, probiotics cannot be recommended in this patient population without further study.

There is little guidance for use of antimicrobial agents to prevent CDI in the oncology patient population, and it is generally not recommended. Various studies have examined the utility of vancomycin or fidaxomicin in preventing CDI. Some studies suggest that use of oral vancomycin may be beneficial as secondary prophylaxis while receiving systemic antibiotic therapy [110, 111]. A new study in the HSCT population suggests that fidaxomicin may have benefit in decreasing CDI for patients receiving fluoroquinolone prophylaxis [112]. Although generally CDI prophylaxis is not recommended, perhaps in certain patient-specific scenarios this should be considered.

The Future of Antimicrobial Stewardship in Oncology and Conclusions

Modern oncology practices should be associated with AMS programs that are in line with hospital-specific oncology policies and based on local epidemiology, drug availability, and hospital- or oncology ward-specific antibiograms. In the future, we might further rely on individualized empiric antibiotic treatment schemes according to a given patient's specific risk factors. An analysis of each oncology division's own AS program with annual assessments of frequency of infections and microbial resistance patterns in this population (particularly the rates of FN), antimicrobial to organism/susceptibility mismatch, adherence to local antibiogram recommendations, allergy assessments, cost savings, antimicrobial related complications, and overall length of stay and mortality assessment should be considered. This may allow for a better local/center-specific/individualized risk assessment pertaining to who should receive antimicrobial prophylaxis for FN while undergoing cytotoxic chemotherapy while further avoiding the development of antimicrobial resistance and antibiotic related complications and preserving and excelling in cancer survival and overall outcomes.

Antimicrobial Stewardship in HIV/AIDS Care

Antimicrobial Resistance and its Impact on People Living with HIV/AIDS

According to most recent data, worldwide, there were approximately 38.4 million people living with HIV (PLWH) in 2021 [113]. In 2020, 30,635 people were diagnosed with HIV, and 1,189,700 people were living with HIV in the USA in 2019 [114]. Around 60% of PLWH in the USA have achieved viral suppression [115], and a large, multisite study in the USA (2001–2008) found that the most common cause of a hospitalization in this population was infection, with bacterial pneumonia, cellulitis, and sepsis topping the list, but also a significant proportion of AIDS-related illnesses [116]. Another large study with 28,057 adult patients examined hospitalization rates and causes among persons with HIV in the USA and Canada from 2005 to 2015 and found that 27% of patients were hospitalized at least once and that 25% of the hospitalizations were non-AIDS defining infections [117]. The most common cause of the infections was sepsis/bacteremia, bacterial pneumonia, and cellulitis/cutaneous abscess, very similar to the prior study [117]. Global estimates of mortality associated with antimicrobial resistance (AMR) in 2019 were 4.95 million, with lower respiratory tract infections being the most burdensome [118]. In contrast, the number of deaths from HIV/AIDS-related illnesses in 2021 was estimated at 650,000 [113]. Life expectancy in PLWH has increased dramatically in those patients treated with antiretroviral therapy (ART), particularly in high-income countries [119, 120]. These additional years of life are likely to come with additional exposure to both antiretroviral therapy and other antimicrobials. There is limited knowledge and statistics regarding antibiotic use in PLWHA or AIDS-related deaths due to infections by antimicrobial resistant infections. A systematic review and meta-analysis of the association of antimicrobial resistance and HIV found 92 low-moderate quality studies between 1995 and 2000 [121] and note that more studies are needed, but their review showed that PLWHA show an increased risk of bacterial AMR. What is known is that many, if not most, of the AIDS-related illnesses are infectious, and are often due to bacteria that commonly appear on the list of those harboring significant resistance. Examples include tuberculosis (TB) and salmonella, but also agents of bacterial pneumonia, such as *Streptococcus pneumoniae* or methicillin-resistant *Staphylococcus aureus*. The CDC report in 2019, called "Antibiotic Resistance Threats in the United States, 2019," listed all abovementioned bacteria, in addition to many more which persons living with HIV/AIDS are more often affected by [122]. Antimicrobial resistance is one of the greatest challenges of our time; it clearly poses a threat to modern medicine and public and global health, and is highly relevant to the population living with HIV/AIDS. The Association for Professionals in Infection Control and Epidemiology defines antimicrobial stewardship as "coordinated interventions designed to improve and measure the appropriate use of antimicrobial agents by promoting the selection of the optimal antimicrobial drug regimen including dosing, duration of therapy, and route of administration" [123] Antimicrobial stewardship strategies and efforts to reduce antimicrobial resistance, both in the hospital setting and in the broader community, in those living with HIV/AIDS are needed; moreover, AMS research regarding this particular population is lacking.

Highlights of Some Antimicrobial Resistant Bacterial Organisms Threats Particularly Relevant to Persons Living with HIV/AIDS

Mycobacterium tuberculosis

PLWH have 20 times greater risk of developing active TB than those without HIV, and globally, 820,000PLWH developed tuberculosis (TB) in 2019 [124]. TB remains the leading cause of death in this population [125]. Nearly 60% of people living with HIV and TB did not know they were co-infected and therefore were not treated [125]. Early recognition, diagnosis, and treatment of TB can have a significant impact on these epidemics. The stewardship concept referred to as the "4 Ds" [126] (the right drug, at the right dose, for the correct duration, and appropriate (de)escalation) is highly relevant in TB. This seems so simple; however; it may not be so straightforward to implement where rapid diagnostics are lacking, most laboratories are not equipped for TB resistance testing in areas of high disease burden, timely diagnosis is often a problem, coordination of care and linkage to care are highly inadequate, and the drugs for treatment may not be consistently available or suffer supply issues [127], or might even be counterfeit and inactive. In addition, in many areas of the world it may be difficult to develop a stewardship program if no denominator of baseline utilization of antimicrobials, antiretroviral medication, or antitubercular drugs has been determined [128]. Few institutions in areas of high endemic TB rates have implemented specific anti-tubercular stewardship programs (ATTSP) that identified critical errors including inaccurate diagnosis and various medication errors; such an implementation can lead to significantly improved outcomes and a decrease in development of resistance [129, 130] and can be an added support in cases of TB with drug resistance [131].

Treatment for drug-susceptible tuberculosis has for decades consisted of the same 4-drug treatment regimen developed in the 1960s—isoniazid, rifampin, pyrazinamide, and ethambutol. This regimen has proved difficult for many reasons. First, it is a complex regimen that tends to interact with many HIV treatment medications and carries a general risk of immune reconstitution syndrome (IRIS). Second, the optimal duration to treat drug-susceptible TB in people living with PLWHA is not known. A new shorter course regimen consisting of 4 months of rifapentine, moxifloxacin, isoniazid, and pyrazinamide was recently added to CDC's recommendations for people with drug-susceptible TB, including PLWHA with CD4 counts at or above 100 uL [132]. Even with the newer shorter duration recommended for eligible people, TB treatment in general remains much longer than most other known bacterial infections. The various causes of acquiring drug-resistant tuberculosis are likely facilitated by the need for long treatment duration and include incomplete or incorrect drug treatment, drug supply issues and drug shortages, counterfeit drugs, and incorrect duration of treatment. Overall community utilization of antimicrobials, including fluoroquinolones, is a factor in the development of resistance [133, 134]. In addition, lack of infection control and nosocomial, shelter, prison, and community spread of drug-resistant TB have been the cause of several recent outbreaks. Multidrug-resistant TB (MDR-TB) is defined as resistance to isoniazid and rifampin, and extensively drug-resistant TB (XDR-TB) is resistance to isoniazid and rifampin plus resistance to any of the fluoroquinolones and at least one injectable second-line aminoglycoside [135]. Treatment of resistant disease may take months to years, has many serious side effects, and is very expensive. In the USA in 2020 there were 7174 cases of tuberculosis [136], in which 56 of these cases were primary MDR-TB and 1 was XDR-TB [137]. Worldwide MDR-TB is a public health crisis and in 2020 only 1 in 3 people with this disease accessed treatment [138]. The average direct cost of treating a case of drug-susceptible TB was 18,000 dollars, whereas treatment for MDR-TB cost 513,000 dollars, and it cost 526,000 dollars to treat one case of XDR-TB [135]. Of note, these figures do not take into account all the indirect expenses.

Applying the concepts of AMS of the 4 Ds and adding the experience from established and successful AMS programs integrated into the TB sphere could add benefit to the many strategies geared toward TB elimination. In addition, the HIV and TB elimination joint effort strategies must be strengthened as these infections often go hand in hand. If TB is to be eliminated, then likely a rational stewardship program needs be applied as a component of an overall strategy [139] to combat this infection. Improved case finding and disease recognition are fundamental to developing any particular strategy aimed at eradicating TB. To this end, rapid diagnostic assays capable of not only identifying TB but also drug-resistant strains are of particular utility. Furthermore, increasing availability of such rapid diagnostics in areas of high disease burden would certainly increase case detection. Identification of cases aside, improvements in the TB treatment cascade itself offer an opportunity to apply AMS concepts. Linkage to care, availability of medications (impacted by drug shortages, interruptions in supply, and prohibitive costs), and appropriate rational therapeutic use (the 4Ds) remain key areas of TB treatment stewardship. An additional thought regarding TB therapy is the concept of applying shorter duration therapies for equivalent efficacy. Novel medications [140] and pathways accelerating drug discovery [141, 142] might advance the protocolization of individualized treatment plans and targeting patient-specific TB strains. TB-specific stewardship programs might be most effective if integrated into existing AMS programs. Furthermore, all health sectors (private and public) should be engaged in TB research, therapy, and care. Antimicrobial, anti-tubercular, and antiretroviral utilization surveillance and control might further the evolution of rapid and efficient outbreak containment. Infection control principles and practices, general measures of reduction of poverty, inequality, and inequity, and preventive measures (BCG vaccination) are critical adjuncts to AMS programs. Additional avenues of enquiry relate to treatment of latent TB in order to prevent development of active tuberculosis, the development of a novel TB vaccine, and doubling efforts to eliminate TB and HIV.

Neisseria gonorrhea

Men who have sex with men (MSM) are disproportionately affected by HIV, and 70% of the new HIV infections in the USA were among gay and bisexual men [143]. The USA has high and ever rising rates of gonorrhea, both in male and female populations, but data suggest the largest increase is seen in MSM [144]. In 2020, 677,769 cases of gonorrhea (45% increase from 2016) were diagnosed in the USA, and antimicrobial resistance is an urgent threat due to the development of significant resistance including ciprofloxacin and ceftriaxone [145, 146]. It was noted that 14% of gonorrhea cases were resistant in 2006 which led to changes in treatment recommendations to avoid ciprofloxacin [146]. The current 2021 updated treatment recommendation for gonorrhea is 500 mg of ceftriaxone given intramuscularly, a recent change from the prior 250 due to concerns for developing resistance [147]. As decreased susceptibility to ceftriaxone is expected to continue, it is essential that surveillance monitoring of minimum inhibitory concentration (MIC) continues and this is being monitored by the Gonococcal Isolate Surveillance Project (GISP) [148]. Gonorrhea cases are being tracked by both the STD surveillance network and GISP in the USA [149].

How can AMS play a role in reducing resistant gonorrhea? First, patients at risk must be tested. Second, because when patients are tested for gonorrhea with nucleic acid amplification which does not test for antimicrobial resistance, and because only patients that fail therapy and return for care are typically re-evaluated by culture with susceptibility testing (an elaborate process requiring special culture media), the development of a rapid test with drug susceptibility analysis could assist in directing specific therapy early [150, 151]. Whole genome testing is another interesting tool to assess for resistance and detect outbreaks [152], similarly to current occasional usage of such testing in tuberculosis outbreaks [153]. Third, using the stewardship principles of the right drug, at the right dose, for the right duration are essential for gonorrhea. Given the increasing resistance to current drugs, development of new medications and/or a vaccine for gonorrhea is needed to combat the serious threat of dwindling treatment options [154–156]. A publication by Xiang-Sheng Chen and colleagues outlined "A ROADMAP Plan to Address Research Needs for Gonococcal Antimicrobial Resistance in China" [157] (with ROADMAP being an abbreviation for "resistance surveillance, outcomes due to AMR, antibiotic stewardship and application, diagnostic tools, mechanisms of AMR, antimicrobial assessment, and population pharmacokinetics"). This scheme can largely be applied to other countries and areas as well. Lastly, global surveillance and tracking is necessary for a better understanding of the problem, and an integrated analysis of patients living with HIV/AIDS and gonorrhea is needed.

Salmonella spp.

Salmonella is one of the most common causes of diarrhea worldwide, and *S. typhi* affects approximately 11–20 million people annually [158]. *Salmonella spp.* can cause infections via contaminated water or food, is primarily of animal origin, and most commonly results in a self-limited enterocolitis [159]. Non-typhoid *Salmonella* is among the most common causes of bacteremia in adults and children in Sub-Saharan Africa, and especially in persons living with HIV [160]. In the USA, it is estimated that about 1.35 million infections a year are caused by non-typhoid Salmonella, and about 5700 infections are due to *S. typhi* infections [160–162]. Salmonella can cause invasive disease with bacteremia as well as disseminated disease (bone, joint, brain) [163]. Invasive Salmonella disease was seen in 8% of the cases reported in USA [163], and invasive disease due to non-typhoid *Salmonella* may be as frequent as 50% in those HIV patients affected [164]. People living with HIV/AIDS with CD4 counts lower than 200 have the highest risk of infection, and although antiretroviral treatment (ART) can reduce severity, frequency, and recurrence, it has been noted that an increased risk persists in people receiving ARVs [165]. In general, treatment for *Salmonella* infections does not differ between those living with HIV/AIDS and people not infected with HIV/AIDS.

In 2013, the USA listed both *S. typhi* and non-typhoid *Salmonella* antimicrobial resistance a serious threat which remains in the newest report from 2019 [122, 162]. In 2015, a total of 8.3% all *S. typhi* infections in the USA were resistant to ciprofloxacin, and 0.3% were resistant to azithromycin; however, there have not been any cases resistant to ceftriaxone in the period between 2009 and 2015 [166]. A total of 2.8% of the non-typhoid *Salmonella* infections in the USA were resistant to ceftriaxone in 2015, 0.4% were resistant to ciprofloxacin, and 0.3% were resistant to azithromycin [166]. The resistance of non-typhoid *Salmonella* to ciprofloxacin is rising and approaching 10% [166]. *S. typhi* was resistant to trimethoprim-sulfamethoxazole (TMP-SMX) in 11.8% of cases in 2015 and 2.4% of cases of non-typhoid Salmonella [166]. This is particularly important with regard to patients living with HIV who are receiving TMP-SMX prophylaxis to prevent *Pneumocystis jirovecii* pneumonia (PJP), as TMP-SMX may not protect against drug-resistant *Salmonella*, thus posing a significant risk.

Internationally, drug-resistant *Salmonella* is much more common. An outbreak of an extensively drug-resistant strain of *S. typhi* including travelers to and from Pakistan [167] highlights the global approach needed to reduce AMR and save lives. A study defining multidrug-resistant (MDR) *Salmonella* as the presence of resistance genes for at least one agent in all three antimicrobial categories of ampicillin/amoxicillin, chloramphenicol, and trimethoprim-sulfamethoxazole found that from 11 Sub-Saharan countries, a total of 50% of all *S. typhi* cases were MDR, and nearly 16% were non-susceptible to a fluoroquinolone [168]. People living with HIV/AIDS using TMP-SMX for prophylaxis for PJP in these areas cannot rely on this antibiotic to protect them against *Salmonella* infection. Children below 15 years of age were more commonly infected with *Salmonella*, and MDR-*Salmonella* was more frequent [169].

S. typhi was listed by World Health Organization (WHO) as high on the priority list for research and development of new antibiotics for antibiotic resistant bacteria in 2018 [170]. The stewardship principle of the right drug, at the right dose, for the right duration, and with the right (de)escalation is essential in *Salmonella* infections, having the potential to prevent morbidity and mortality, especially in Africa where the risk of MDR-*Salmonella* is particularly high and where the burden of HIV/AIDS is greatest. Prevention of *S. typhi* disease by vaccination (purified antigen injectable vaccine or a live-attenuated oral vaccine) is a well-known approach applied to travelers to endemic areas, and the WHO recently approved a typhoid conjugate vaccine with longer lasting immunity with a primary focus on citizens living in typhoid endemic countries [168]. Another essential preventive intervention is good handwashing. This is a cheap method of prevention, but is not always accessible where the need is highest and where many immunocompromised persons live with HIV/AIDS [171]. According to the UN, 2.4 billion people do not have access to toilets/latrines, and 3/10 of people do not have access to clean drinking water [172]. Food safety practices targeted to people living with HIV/AIDS advise to cook food properly and to promptly refrigerate leftovers and meats [173]. Those recommendations may be followed to prevent foodborne illnesses such as *Salmonella* for most people in the USA, as 99.9% of American households own a refrigerator, and 99.7% own cooking appliances [174]. However, a statistic from Euromonitor in 2015 notes only a total of 76% average refrigerator ownership globally, and possibly even less when considering that about 15% of the world's population lacks access to reliable electricity [175]. Working toward global reduction in poverty, increased access to clean water and sanitation, and working toward health equity will decrease *Salmonella* infections. Finally, by using the One Health Approach including working to improve the health of people, animals, and the environment [176], a reduction of overall *Salmonella* infection and *Salmonella* resistance might be achieved.

Streptococcus pneumoniae

S. pneumoniae causes more than two million disease episodes in the USA yearly, and it is the most common cause of pneumonia and meningitis in the USA [122]. People living with HIV/AIDS are at a significantly increased risk of *S. pneumoniae* pneumonia, but also are at increased risk of complications such as pneumococcal bacteremia, as well as higher rates of antimicrobial resistance [177]. There is no method known to prevent exposure to *S. pneumoniae*, but vaccination is a well-documented means of preventing streptococcal invasive disease and reduces all-cause mortality [178]. In within the United States, people living with HIV/AIDS are recommended to receive pneumococcal conjugate vaccine 20 (PCV20) or pneumococcal conjugate vaccine 15 (PCV15) followed by pneumococcal polysaccharide vaccine (PPSV23) [177, 178]. The CDC recorded *Streptococcus pneumoniae* resistance as a serious threat in 2013 and stated that it required prompt and sustained action to prevent the problem from growing [121]. Soon after the introduction of antibiotics, *Streptococcus* began to develop resistant clones that not only widely disseminated worldwide, but had also been found difficult to control due to selective antibiotic pressure, and moreover, it is a commensal in the upper respiratory tract and shares genes with other commensal bacteria in this location [179]. Penicillin resistance was discovered in 1965, resistance to macrolides and trimethoprim-sulfamethoxazole began rising in the 1970–1980s, and then resistance to tetracyclines, clindamycin, and fluoroquinolones developed in quick succession [179], culminating in ceftriaxone resistance [180]. Macrolide and penicillin group resistance is common in the USA, and one in three serious *S. pneumoniae* infections is resistant to one or more antibiotics [122]. A study from Spain described patients living with HIV/AIDS on TMP-SMX prophylaxis for PJP as more likely to present with TMP-SMX-resistant pneumococcal disease, and with a trend toward penicillin resistance [181]. Another study from France showed that the use of TMP-SMX increased the likelihood of penicillin-resistant *Streptococcus pneumoniae* [182]. In malaria endemic areas in Africa where treatment with sulfadoxine-pyrimethamine is common, an increased resistance of pneumococcus resistance to TMP-SMX has been seen [183, 184]. Antimicrobial stewardship efforts centered around the 4 D's and antimicrobial use only when it is clearly indicated, are essential for the prevention of pneumococcal resistance. It is important to be aware of the "collateral damage effect" of antibiotics and attempt to reduce unnecessary use of PJP prophylaxis and sulfa-based malaria treatment as these have a significant negative impact on pneumococcal resistance (and likely other bacteria as well). Vaccination for all would significantly reduce pneumococcal infections and overall antibiotic use. The WHO noted that "if every child in the world received a vaccine to protect them from infection with *Streptococcus pneumoniae* bacteria, this would prevent an estimated 11 million days of antibiotic use each year" [185].

Antiretroviral Therapy Stewardship

Antiretroviral (ARV) stewardship [186–189] is a strategy to improve overall clinical outcomes in HIV care. Several studies have evaluated stewardship interventions on antiretroviral therapy in both outpatient and inpatient settings, including effects on medication errors, adherence, drug interactions, antiretroviral resistance, cost, and clinical outcomes [190–193]. Many of these interventions involved a pharmacist trained in infectious diseases, and these studies have all shown a positive effect and increased patient safety. Healthcare facilities should consider developing ARV stewardship programs [189].

Data from the START trial demonstrated that immediate ARV therapy reduced the incidence of several severe bacterial infections in people living with HIV with higher CD4 counts [194]. This effect was largely mediated by an increase in CD4 counts. We postulate that ARV stewardship, by increasing CD4+ counts and improving clinical outcomes, can lead to a decrease in bacterial infections and consequently an overall reduction in antimicrobial utilization leading to less antimicrobial resistance.

Antimicrobial Stewardship Efforts Regarding People Living with HIV/AIDS to Assist in Overall Reduction of Antimicrobial Resistance

- Encourage being up to date with currently recommended vaccines that prevent infections, superinfections, and resistant microbes, thus contributing to the reduction of global consumption of antibiotics.
- Initiate ARV early after diagnosis to improve retention in care and to increase sustained viral suppression [194–196]. Overall, this may prevent AIDS and in turn opportunistic infections, decrease the incidence of severe bacterial infections, and reduce the need for antimicrobials and therefore antimicrobial resistance.
- Ensure that TMP-SMX prophylaxis of *Pneumocystis jirovecii* pneumonia is prescribed according to current guidelines to avoid emergence of antimicrobial resistant organisms.
- Develop ARV stewardship programs in healthcare facilities.

Antimicrobial Stewardship in HIV/AIDS Care Conclusions

Antimicrobial stewardship in HIV/AIDS care is complex and can be very challenging. To reduce overall antimicrobial resistance and antimicrobial resistant infections which often affect this vulnerable population, it is essential to devote more research funding to this topic, develop research studies that involve people living with HIV/AIDS affected by these infections, and develop AMS programs and implement AMS practices into standard AIDS care worldwide.

Antimicrobial Stewardship in Solid Organ Transplant (SOT) and Hematopoietic Stem Cell Transplant (HSCT) Programs

While AMS programs are mandated at the institutional level by the Center for Medicare and Medicaid Services (CMS), there is no rule stating that transplant programs must come under the stewardship umbrella [197]. In a 2016 voluntary survey of transplant centers across the USA, only 74% had AMS programs that were involved in transplant-related interventions and activities [198]. Moreover, there is little data on antimicrobial selection, duration, and de-escalation in the transplant population, despite the increased risk of multidrug-resistant organism (MDRO) infections in these patients. The complexity of factors associated with transplant recipients' infection risk, which includes type of transplant, net state of immunosuppression, and characteristics of episodes of rejection, makes it challenging to attempt to formulate comprehensive guidelines. At the same time, there is an urgent need to develop AMS interventions in the transplant population.

Epidemiology of Antimicrobial Resistance in SOT and HSCT Recipients

Infection with MRSA remains a leading cause of morbidity and mortality in SOT. In a meta-analysis, the pooled prevalence of MRSA colonization in liver, kidney, and lung recipients was 9.4% (95% CI 3.0–18.5%) post-transplantation compared to 2% in the general population [199]. In addition to the traditional risk factors for MRSA infections, unique organ-specific risk factors have been well identified. In liver transplant recipients, risk factors include alcoholic cirrhosis, primary CMV infection, and prolonged operative time [200, 201]. In contrast, the major risk MRSA factor for lung transplant recipients is ventilation for more than 5 days [202].

Infections with MDRO including carbapenem-resistant Enterobacterales (CRE) and carbapenem-resistant *Acinetobacter baumannii (CRAB)* have been on the rise and pose one of the biggest antimicrobial resistance threats in this century [166]. Transplant recipients are at increased risk of such infection given frequent exposures to the healthcare system and high burden of antimicrobial use both before and after transplantation. In a 2020 review, the prevalence of CRE ranged between 5% and 27% post-transplantation [203]. Infections with CRE are often related to the transplanted organ. The strongest risk factor for infection is prior colonization with CRE, which highlights the importance of prevention [204, 205].

HSCT recipients are at risk for MDRO including CRE as well. They are a unique population in that they often receive prolonged antimicrobial prophylaxis, usually with fluoroquinolones, while also having protracted neutropenia. Predisposing factors for CRE in this patient population include allogenic HSCT, breakthrough bacteremia while on carbapenems, and longer duration of hospitalization [206].

Antimicrobial Stewardship Challenges in HSCT and SOT Programs

One of the mainstays of AMS is to reduce the unnecessary use of broad-spectrum antimicrobials. This is done by evaluating the patient's risk factors for resistance and targeting antimicrobials to the most likely site of infection. For example, for skin and soft tissue infections, antibiotics should cover Gram-positive skin flora only; for intra-abdominal infections, antifungal coverage should be avoided unless a culture is positive for yeast. Transplant recipients pose an exceptional challenge to the steward, as it is not uncommon for patients to have extensive histories of hospitalizations and infections prior to transplantation, associated with frequent receipt of antibiotics.

As such, transplant recipients presenting with suspected or confirmed infection often require the use of broad-spectrum agents. Inadequate or delayed antimicrobial coverage has been shown to lead to devastating outcomes [207–209], and these patients' suppressed immune systems leave no room for error. Thus, a clinician who might otherwise consider using a narrower spectrum antibiotic in another clinical setting may be wary of "getting it wrong" and putting the patient's life at risk.

Another primary tenet of stewardship is de-escalation of antimicrobials based on culture results. However, it can be a daunting task for the steward to convince transplant care teams to stop or change an antibiotic regimen to which a patient has responded well and quickly. It can be even more challenging to narrow the regimen after identifying an organism when the patient is critically ill. Pushback is common, despite the increasing evidence that appropriate de-escalation is not associated with increased adverse events; on the contrary, it appears to be associated with decreased risk of *Clostridium difficile* infection and shorter hospital stays [210, 211].

Quality metrics and process standards for AMS are well established by the CDC and the Joint Commission in the USA. Recently, the American Society of Transplantation Infectious Disease Community of Practice published a series of guidelines that established a framework for transplant infectious disease physicians [212]. However, for many common clinical scenarios, a paucity of rigorously conducted research studies hinders the ability to generate evidence-based guidelines. Furthermore, the complexity of these patients' conditions, heterogeneity of their clinical presentation, and variability in immunosuppression combined with long-held beliefs by transplant care providers about how to prescribe antibiotics make implementing any new guidelines especially difficult.

Furthermore, new technologies in transplantation development outpace the formulation of guidelines. Nowhere is this more apparent than in the world of heart failure management and cardiac transplantation. As an example, the use of durable (ventricular assist devices that patients go home with) and temporary (generally limited to intensive care unit use) mechanical circulatory support (MCS) as a bridge to cardiac transplant has exploded in recent years. Up to 40% of patients are bridged with durable MCS prior to heart transplantation [213]. Specifically, over 25,000 patients underwent left ventricular assist device (LVAD) implantation from 2010 to 2019, with 41% of these patients developing an infection in the first year post-implantation [214]. The 1-year mortality for patients with LVAD infection was 29% compared to 13% in patients without infection [215]. Despite how common the occurrence, there is a paucity of evidence to determine the appropriate duration of antimicrobial therapy for retained MCS and other devices in patients awaiting cardiac transplant. Once an infection develops, some specialists continue therapy until the device is removed at the time of transplant, which could be many months or even years, while others define a course of weeks or months and then if clinically resolved, discontinue and observe. Frequent infections and long courses of antibiotics mean that by the time a patient receives a cardiac transplant, he or she is likely to be heavily colonized by resistant organisms. When combined with aggressive immunosuppression, this provides a setup for dangerous infections. Moreover, there are no national guidelines regarding which agents to use for peri-operative surgical prophylaxis (for MCS device implantation or cardiac transplantation) or whether prophylactic antibiotics or antifungal agents are needed throughout the period of bridging with temporary MCS, which is known to be high risk for infection. This creates significant variability in practice between transplant centers and in some instances even between surgeons in the same center.

Additionally, advances in the field of xenotransplantation are anticipated to add new challenges to transplant ID practice [216]. The potential transmission of infectious agents from nonhuman species to human recipients is uncharted territory [217].

Opportunities to Optimize Antimicrobial Therapy for HSCT and SOT: Leveraging Existing Outlines and Frameworks

It is critical that we begin to move toward the utilization of existing concepts and frameworks such as the CDC's elements of AMS and handshake stewardship with modifications to fit within a transplant stewardship program [218, 219]. A 2019 review suggests that publications related to stewardship in HSCT programs are thankfully increasing [210]. The review included studies with varied stewardship interventions including allergy assessments, implementation of clinical pathways, and early antimicrobial de-escalation. Overall, the review showed a decrease in antibiotic use with no adverse impact on length of stay or mortality. For instance, in a single-center retrospective study, early de-escalation of antimicrobials to prophylaxis in allogenic HSCT with neutropenic fever before engraftment led to significant reductions in Gram-positive broad-spectrum agents without impacting the rate of fever recurrence, ICU admission, or in-hospital mortality [220]. In another successful example from a SOT program, an academic center in Canada implemented twice weekly detailing rounds with transplant pharmacists and prescribers caring for SOT patients from 2013 to 2016. The recommendations applied AMS principles and accounted for the patients' clinical condition as a whole. Concordance with AMS recommendations increased from 69% (60/87) to 83.7% (93/111), $p = 0.01$ [221]. The evaluation of antibiograms specific to transplant populations can aid in the selection of empiric antimicrobial coverage [222]. Early infectious disease consultation in SOT recipients has been shown to decrease mortality and reduce rehospitalization rates [223]. Furthermore, targeted antiviral and antifungal interventions could improve the rates of appropriate use [224]. Given the complexity of transplant patient care, emphasizing the importance of audit with feedback, collecting specific local SOT and HSCT data, and sharing the findings with the transplant team could facilitate the adoption of new interventions and ideas.

The rising rates of antimicrobial resistance, the fast pace of transplant technology, and the unique complexities of the transplant population bring an exceptional challenge to AMS teams. The future landscape of AMS in SOT and HSCT should include universal quality standards and measures as well as national and international collaborations between transplant centers.

Acknowledgement A great thank you to Dr. Jose Caro for offering input to the AMS in HIV/AIDS Care chapter.

References

1. Klastersky J, de Naurios J, Rolston K, et al. Management of febrile neutropaenia: ESMO Clinical Practice Guidelines. Ann Oncol. 2016;25(Suppl 5):v111–8.
2. Tai E, Guy GP, Dunbar A, Richardson LC. Cost of cancer-related neutropenia or fever hospitalizations, United States, 2012. J Oncol Pract. 2017;13(6):e552–61.
3. Caggiano V, Weiss RV, Rickert TS, Linde-Zwirble WT. Incidence, cost, and mortality of neutropenia hospitalization associated with chemotherapy. Cancer. 2005;103:1916–24.
4. Freifeld AG, Bow EJ, Sepkowitz KA, Boeckh MJ, Ito JI, Mullen CA, Raad II, Rolston KV, Young JA, Wingard JR, et al. Clinical practice guideline for the use of antimicrobial agents in neutropenic patients with cancer: 2010 update by the Infectious Diseases Society of America. Clin Infect Dis. 2011;52:427–31.
5. Rolston KV. Challenges in the treatment of infections caused by gram-positive and gram-negative bacteria in patients with cancer and neutropenia. Clin Infect Dis. 2005;40(Suppl 4):S246–52.
6. Ohmagari N, Hanna H, Graviss L, et al. Risk factors for infections with multidrug-resistant *Pseudomonas aeruginosa* in patients with cancer. Cancer. 2005;104(1):205–12.
7. Mihu CN, Rhomberg PR, Jones RN, et al. Escherichia coli resistance to quinolones at a comprehensive cancer center. Diagn Microbiol Infect Dis. 2010;67(3):266–9.
8. Rangaraj G, Granwehr BP, Jiang Y, et al. Perils of quinolone exposure in cancer patients. Cancer. 2010;116(4):967–73.
9. Trubiano JA, Worth LJ, Thursky KA, Slavin MA. The prevention and management of infections due to multidrug resistant organisms in haematology patients. Br J Clin Pharmacol. 2015;79(2):195–207.
10. Horasan ES, Ersoz G, Tombak A, Tiftik N, Kaya A. Bloodstream infections and mortality-related factors in febrile neutropenic cancer patients. Med Sci Monit. 2011;17(5):Cr304–9.
11. Montassier E, Batard E, Gastinne T, et al. Recent changes in bacteremia in patients with cancer: a systematic review of epidemiology and antibiotic resistance. Eur J Clin Microbiol Infect Dis. 2013;32:841–50.
12. Mikulska M, Viscoli C, Orasch C, et al. Aetiology and resistance in bacteraemias among adult and paediatric haematology and cancer patients. J Infect. 2014;68(4):321–31.
13. Rice LB. Federal funding for the study of antimicrobial resistance in nosocomial pathogens: no ESKAPE. J Infect Dis. 2008;197:1079–81.
14. Rapoport B, Klastersky J, Raftopoulos H, et al. The emerging problem of bacterial resistance in cancer patients; proceedings of a workshop held by MASCC "neutropenia, infection and myelosuppression" study group during the MASCC annual meeting held in Berlin. Support Care Cancer. 2016;24:2819–26.
15. Ansari SR, Hanna H, Hachem R, et al. Risk factors for infections with multidrug-resistant Stenotrophomonas maltophilia in patients with cancer. Cancer. 2007;109:2615–22.
16. Gudiol C, Catalayud L, Garcia-Vidal C, et al. Bacteraemia due to extended-spectrum beta-lactamase-producing Escherichia coli (ESBL-EC) in cancer patients: clinical features, risk factors, molecular epidemiology and outcome. J Antimicrob Chemother. 2010;65:333–41.
17. Gudiol C, Tubau F, Calatayud L, et al. Bacteraemia due to multi-drug resistant gram-negative bacilli in cancer patients: risk factors, antibiotic therapy and outcomes. J Antimicrob Chemother. 2011;66:657–63.
18. Huh K, Kang CI, Kim J, et al. Risk factors and treatment outcomes of bloodstream infection caused by extended-spectrum cephalosporin-resistant Enterobacter species in adults with cancer. Diagn Microbiol Infect Dis. 2014;78:172–7.
19. Garnica M, Maiolino A, Nucci M. Factors associated with bacteremia due to multidrug-resistant gram-negative bacilli in hematopoietic stem cell transplant recipients. Braz J Med Biol Res. 2009;42:89–93.
20. Oliveira AL, Souza M, Carvalho-Dias VM, et al. Epidemiology of bacteremia and factors associated with multi-drug resistant gram-negative bacteremia in hematopoietic stem cell transplant recipients. Bone Marrow Transplant. 2007;39:775–81.

21. Johnson LE, D'Agata EM, Paterson DL, et al. *Pseudomonas aeruginosa* bacteremia over a 10-year period: multidrug resistance and outcomes in transplant recipients. Transpl Infect Dis. 2009;11:227–43.
22. Kim SH, Kwon JC, Choi SM, et al. Escherichia coli and *Klebsiella pneumonia* bacteremia in patients with neutropenia fever: factors associated with extended-spectrum Beta-lactamase production and its impact on outcome. Ann Hematol. 2013;92:533–41.
23. Bodro M, Gudiol C, Garcia-Vidal C, et al. Epidemiology, antibiotic therapy and outcomes of bacteremia caused by drug-resistant ESKAPE pathogens in cancer patients. Support Care Cancer. 2014;22:603–10.
24. Ghanem H, Hachem R, Jiang Y, et al. Outcomes for and risk factors associated with vancomycin-resistant *Enterococcus faecalis* and vancomycin-resistant *Enterococcus faecium* bacteremia in cancer patients. Infect Control Hosp Epidemiol. 2007;28:1054–9.
25. Ariffin H, Navaratnam P, Mohamed M, et al. Ceftazidime-resistant *Klebsiella pneumonia* bloodstream infection in children with febrile neutropenia. Int J Infect Dis. 2000;4:21–5.
26. Rangaraj G, Granwehr BP, Jiang Y, et al. Perils of quinolone exposure in cancer patients: breakthrough bacteremia with multidrug-resistant organisms. Cancer. 2010;116:967–73.
27. Weinstock DM, Conlon M, Iovino C, et al. Colonization, bloodstream infection, and mortality caused by vancomycin-resistant enterococcus early after allogeneic hematopoietic stem cell transplant. Biol Blood Marrow Transplant. 2007;13:615–21.
28. Gudiol C, Bodro M, Simonetti A, et al. Changing aetiology, clinical features, antimicrobial resistance, and outcomes of bloodstream infection in neutropenic cancer patients. Clin Microbiol Infect. 2013;19:474–9.
29. Samonis G, Vardakas KZ, Maraki S, et al. A prospective study of characteristics and outcomes of bacteremia in patients with solid organ or hematologic malignancies. Support Care Cancer. 2013;21:2521–6.
30. Cattaneo C, Antoniazzi F, Casari S, et al. P. aeruginosa bloodstream infections among hematological patients: an old or new question? Ann Hematol. 2012;91:1299–304.
31. Nesher L, Rolston KV. The current spectrum of infection in cancer patients with chemotherapy related neutropenia. Infection. 2014;42:5–13.
32. Satlin MJ, Calfee DP, Chen L, et al. Emergence of carbapenem-resistant enterobacteriaceae as causes of bloodstream infections in patients with hematologic malignancies. Leuk Lymphoma. 2013;54(4):799–806.
33. Fukata Y, Robert R, Muder MD, et al. Risk factors for acquisition of multidrug-resistant Acinetobacter baumannii among cancer patients. Am J Infect Control. 2013;41:1249–52.
34. Sood P, Seth T, Kapil A, et al. Emergence of multidrug resistant Acinetobacter blood stream infections in febrile neutropenia patients with haematological cancers and bone marrow failure syndromes. J Indian Med Assoc. 2012;110:439–44.
35. Irfan S, Idrees F, Mehraj V, et al. Emergence of carbapenem resistant gram negative and vancomycin resistant gram positive organisms in bacteremic isolates of febrile neutropenic patients: a descriptive study. BMC Infect Dis. 2008;8:80.
36. deKraker ME, Jarlier V, Monen JC, et al. The changing epidemiology of bacteremia in Europe: trend from the European Antimicrobial Resistance Surveillance System. Clin Microbiol Infect. 2013;19:860–8.
37. Kallen AJ, Mu Y, Bulens S, et al. Heath care associated invasive MRSA infections, 2005-2008. JAMA. 2010;304:641–8.
38. Linder KL, Nicolau DP, Nailor MD. Predicting and preventing antimicrobial resistance utilizing pharmacodynamics: part I gram positive bacteria. Expert Opin Drug Metab Toxicol. 2016;12(3):267–80.
39. Doern GV, Ferraro MJ, Brueggemann AB, et al. Emergency of high rates of antimicrobial resistance among viridans group streptococci in the United States. Antimicrob Agents Chemother. 2001;47(1):87–91.
40. Marron A, Carratala J, Fet A, et al. High rates of resistance to cephalosporins among viridans-group streptococci causing bacteraemia in neutropenic cancer patients. J Antimicrob Chemother. 2001;47(1):87–91.

41. Bruckner L, Gigliotti F. Viridans group streptococcal infections among children with cancer and the importance of emerging antibiotic resistance. Semin Pediatr Infect Dis. 2006;17(3):153–60.
42. Yim J, Smith JR, Rybak J. Role of combination antimicrobial therapy for vancomycin-resistant Enterococcus faecium infections: review of the current evidence. Pharmacotherapy. 2017;37(5):579–92.
43. Barlam TF, Cosgrove SE, Abbo LM, et al. Implementing an antibiotic stewardship program: guidelines by the Infectious Diseases Society of America and the Society for Healthcare Epidemiology of America. Clin Infect Dis. 2016;62(10):e51.
44. Spellberg B. The new antibiotic mantra—"shorter is better". JAMA Intern Med. 2016;176(9):1254–5.
45. Rolston KVI, Bodey GP, Safdar A. Polymicrobial infection in patients with cancer: an underappreciated and underreported entity. Clin Infect Dis. 2007;45(2):228–33.
46. Goff DA, Kullar R, Bauer KA, File TM. Eight habits of highly effective antimicrobial stewardship programs to meet the joint commission standards for hospitals. Clin Infect Dis. 2017;64(8):1134–9.
47. Commonwealth of Massachusetts. Massachusetts antibiograms. https://www.mass.gov/service-details/massachusetts-antibiograms. Accessed 4 Apr 2019.
48. Overly S, Hayes S, Mehta J, Hamilton K, Peterson D. Evaluating patient-specific antibiograms. Open Forum Infect Dis. 2017;4(Suppl 1):S258.
49. Huang K-HG, Cluzet V, Hamilton K, Fadugba O. The impact of reported beta-lactam allergy in hospitalized patients with hematologic malignancies requiring antibiotics. Clin Infect Dis. 2018;67(1):27–33.
50. Artau A, Taremi M, Foolad F, et al. Safety, efficacy and clinical impact of penicillin allergy skin testing in immunocompromised cancer patients at a comprehensive cancer center. Poster. ID Week 2018. https://idsa.confex.com/idsa/2018/webprogram/Paper69777.html. Accessed 10 Apr 2019.
51. National Comprehensive Cancer Network. Prevention and treatment of cancer-related infections (version 1.2019). https://www.nccn.org/professionals/physician_gls/pdf/infections.pdf. Accessed June 2022.
52. Le Clech L, Talarmin JP, Couturier MA, et al. Early discontinuation of empirical antibacterial therapy in febrile neutropenia: the ANTIBIOSTOP study. Infect Dis. 2018;50(7):539–49.
53. Snyder M, Pasikhova Y, Baluch A. Early antimicrobial de-escalation and stewardship in adult hematopoietic stem cell transplantation recipients: retrospective review. Open Forum Infect Dis. 2017;4(4):226.
54. Aguilar-Guisado M, Espigado I, Martín-Peña A, et al. Optimisation of empirical antimicrobial therapy in patients with haematological malignancies and febrile neutropenia (how long study): an open-label, randomised, controlled phase 4 trial. Lancet Haematol. 2017;4(12):e573–83.
55. Taplitz RA, Kennedy EB, Bow EJ, et al. Antimicrobial prophylaxis for adult patients with cancer-related immunosuppression: ASCO and IDSA clinical practice guideline update. J Clin Oncol. 2018;36(30):3043–54.
56. Horton LE, Haste NM, Taplitz RA. Rethinking antimicrobial prophylaxis in the transplant patient in the world of emerging resistant organisms-where are we today? Curr Hematol Malig Rep. 2018;13(1):59–67.
57. Heidenreich D, Kreil S, Nolte F, et al. Allogeneic hematopoietic cell transplantation without fluconazole and fluoroquinolone prophylaxis. Ann Hematol. 2016;95:287–93.
58. Kern WV, Klose K, Jellen-Ritter AS, et al. Fluoroquinolone resistance of Escherichia coli at a cancer center: epidemiologic evolution and effects of discontinuing prophylactic fluoroquinolone use in neutropenic patients with leukemia. Eur J Clin Microbiol Infect Dis. 2005;24(2):111–8.
59. Chong Y, Yakushiji H, Ito Y, Kamimura T. Clinical impact of fluoroquinolone prophylaxis in neutropenic patients with hematological malignancies. Int J Infect Dis. 2011;15:e277–81.

60. Saito T, et al. Effects on spectrum and susceptibility patterns of isolates causing bloodstream infection by restriction of fluroquinolone prophylaxis in a heamatology-oncology unit. Eur J Clin Microbiol Infect Dis. 2007;27(3):209–16.
61. Carratala J, Fernandez-Sevilla A, Tubau F, Callis M, Gudiol F. Emergence of quinolone-resistant Escherichia coli bacteremia in neutropenic patients with cancer who have receiving prophylactic norfloxacin. Clin Infect Dis. 1995;20:557–60.
62. Cometta A, Calandra T, Bille J, Glauser MP. Escherichia coli resistant to fluoroquinolones in patients with cancer and neutropenia. N Engl J Med. 1994;330:1240–1.
63. Sánchez-Díaz AM, Cuartero C, Rodríguez JD, et al. The rise of ampicillin-resistant Enterococcus faecium high-risk clones as a frequent intestinal colonizer in oncohaematological neutropenic patients on levofloxacin prophylaxis: a risk for bacteraemia? Clin Microbiol Infect. 2016;22(1):59.e1–8.
64. Gudiol C, Ayats J, Camoez M, et al. Increase in bloodstream infection due to vancomycin-susceptible *Enterococcus faecium* in cancer patients: risk factors, molecular epidemiology and outcomes. PLoS One. 2013;8:e74734.
65. Weber D, Hiergeist A, Weber M, et al. Detrimental effect of broad-spectrum antibiotics on intestinal microbiome diversity in patients after allogeneic stem cell transplantation: lack of commensal sparing antibiotics. Clin Infect Dis. 2019;68(8):1303–10.
66. Scappaticci GB, Perissinotti AJ, Nagel JL, et al. Risk factors and impact of *Clostridium difficile* recurrence of haematology patients. J Antimicrob Chemother. 2017;72(5):1488–95.
67. Chung MS, Kim J, Kang JO, et al. Impact of malignancy on *Clostridium difficile* infection. Eur J Clin Microbiol Infect Dis. 2016;35(11):1771–6.
68. Tai E, Richardson LC, Townsend J, et al. *Clostridium difficile* infection among children with cancer. Pediatr Infect Dis J. 2011;30(7):610–2.
69. Chopra T, Chandrasekar P, Salimnia H, et al. Recent epidemiology of *Clostridium difficile* infection during hematopoietic stem cell transplantation. Clin Transpl. 2011;25(1):E82–7.
70. Delgato A, Reveles I, Cabello F, et al. Poorer outcomes among cancer patients diagnosed with *Clostridium difficile* infections in United States community hospitals. BMC Infect Dis. 2017;17:448.
71. Campbell R, Dean B, Nathanson B, Haidar T, Strauss M, Thomas S. Length of stay and hospital costs among high-risk patients with hospital-origin *Clostridium difficile*-associated diarrhea. J Med Econ. 2013;16(3):440–8.
72. Yoon YK, Kim MJ, Sohn JW, et al. Predictors of mortality attributable to *Clostridium difficile* infection in patients with underlying malignancy. Support Care Cancer. 2014;22(8):2039–48.
73. Wang MS, Evans CT, Rodriguez T, Gerding DN, Johnson S. *Clostridium difficile* infection and limitations of markers for severity in patients with hematologic malignancy. Infect Control Hosp Epidemiol. 2013;34(2):127–32.
74. Apostolopoulou E, Raftopoulos V, Terzis K, Elefsiniotis I. Infection probability score: a predictor of *Clostridium difficile*-associated disease onset in patients with haematological malignancy. Eur J Oncol Nurs. 2011;15(5):404–9.
75. Donskey CJ, Kundrapu S, Deshpande A. Colonization versus carriage of *Clostridium difficile*. Infect Dis Clin N Am. 2015;29:13–28.
76. Bartlett JG, Chang TW, Gurwith M, et al. Antibiotic-associated pseudomembranous colitis due to toxin producing clostridia. N Engl J Med. 1978;298:531–4.
77. Kamthan AG, Bruckner HW, Hirschman SZ, et al. *Clostridium difficile* diarrhea induced by cancer chemotherapy. Arch Intern Med. 1992;152(8):1715–7.
78. Thielman NM, Wilson KH. Chapter 96: antibiotic-associated colitis. In: Mandell, Douglas, and Bennett's principles and practice of infectious diseases. 7th ed.; 2004. p. 1375–1384.
79. Hebbard A, Slavin MA, Reed C, et al. The epidemiology of *Clostridium difficile* infection in patients with cancer. Expert Rev Anti-Infect Ther. 2016;14(11):1077–85.
80. Anand A, Glatt AE. *Clostridium difficile* infection associated with antineoplastic chemotherapy: a review. Clin Infect Dis. 1993;17:109–13.
81. Morales Chamorro R, Serrano Blanch R, Méndez Vidal MJ, et al. Pseudomembranous colitis associated with chemotherapy with 5-fluorouracil. Clin Transl Oncol. 2005;7:258–61.

82. Bilgrami S, Feingold JM, Dorsky D, et al. Incidence and outcome of *Clostridium difficile* infection following autologous peripheral blood stem cell transplantation. Bone Marrow Transplant. 1999;23:1039–42.
83. Gorschlüter M, Glasmacher A, Hahn C, et al. *Clostridium difficile* infection in patients with neutropenia. Clin Infect Dis. 2001;33:786–91.
84. Nielson H, Daugaard G, Tvede M, et al. High prevalence of *Clostridium difficile* diarrhea during intensive chemotherapy for disseminated germ cell cancer. Br J Cancer. 1992;66(4):666–7.
85. Emoto M, Kawarabayashi T, Hachisuga MD. *Clostridium difficile* colitis associated with cisplatin-based chemotherapy in ovarian cancer patients. Gynecol Oncol. 1996;61(3):369–72.
86. Husain A, Aptaker L, Spriggs DR, et al. Gastrointestinal toxicity and *Clostridium difficile* diarrhea in patients treated with paclitaxel-containing chemotherapy regimens. Gynecol Oncol. 1998;71(1):104–7.
87. Loo VG, Poirier L, Miller MA, et al. A predominantly clonal multi-institutional outbreak of *Clostridium difficile*-associated diarrhea with high morbidity and mortality. N Engl J Med. 2005;353:2442–9.
88. Hautmann MG, Hipp M, Kolbl O. *Clostridium difficile*-associated diarrhea in radiooncology: an underestimated problem for the feasibility of the radiooncological treatment. Radiat Oncol. 2011;6:89.
89. Howell MD, Novack V, Grgurich P, et al. Iatrogenic gastric acid suppression and the risk of nosocomial *Clostridium difficile* infection. Arch Intern Med. 2010;170:784–90.
90. Bishop KD, Castillo JJ. Risk factors associated with *Clostridium difficile* infection in adult oncology patients with a history of recent hospitalization for febrile neutropenia. Leuk Lymphoma. 2012;53:1617–9.
91. McCaleb RV, Gandhi AS, Clark SM, et al. Clinical outcomes of acid suppressive therapy use in hematology/oncology patients at an academic medical center. Ann Pharmacother. 2016;50:541–7.
92. Khan A, Raza S, Batul SA, et al. The evolution of *Clostridium difficile* infection in cancer patients: epidemiology, pathophysiology, and guidelines for prevention and management. Recent Pat Antiinfect Drug Discov. 2012;7(2):157–70.
93. Peretz A, Shlomo IB, Nitzan O, et al. *Clostridium difficile* infection: associations with chemotherapy, radiation therapy, and targeting therapy treatments. Curr Med Chem. 2016;23(39):4442–9.
94. McDonald LC, Gerding DN, Johnson S, et al. Clinical practice guidelines for Clostridium difficile infection in adults and children: 2017 update by the Infectious Diseases Society of America (IDSA) and Society for Healthcare Epidemiology of America (SHEA). Clin Infect Dis. 2018;66(7):e1–e48.
95. Nerandzic MM, Mullane K, Miller MA, et al. Reduced acquisition and overgrowth of vancomycin-resistant enterococci and candida species in patients treatment with fidaxomicin versus vancomycin for *Clostridium difficile* infection. Clin Infect Dis. 2012;55:S121–6.
96. Surawicz CM, Brandt LJ, Binion DG, et al. Guidelines for diagnosis, treatment, and prevention of *Clostridium difficile* infections. Am J Gastroenterol. 2013;108:478–98.
97. Johnson S, Lavergne V, Skinner AM, et al. Clinical practice guideline by the Infectious Disease Society of America (IDSA) and Society for Healthcare Epidemiology of America (SHEA): 2021 focused update guidelines on management of Clostridioides difficile infection in adults. Clin Infect Dis. 2021;73(5):e1029–44. https://doi.org/10.1093/cid/ciab54.
98. Musher DM, Aslam S, Logan N, et al. Relatively poor outcome after treatment of *Clostridium difficile* colitis with metronidazole. Clin Infect Dis. 2005;40:1586–90.
99. Siegfried J, Dubrovskaya Y, Flagiello T, et al. Initial therapy for mild to moderate *Clostridium difficile* infection: exploring the role of oral metronidazole versus vancomycin in 168 hospitalized patients. Infect Dis Clin Pract. 2016;24:210–6.
100. Stevens VW, Nelson RE, Schwab-Daugherty EM, et al. Comparative effectiveness of vancomycin and metronidazole for the prevention of recurrence and death in patients with *Clostridium difficile* infection. JAMA Intern Med. 2017;144(4):546–53.

101. Cornely OA, Ma M, Fantin B, et al. Resolution of *Clostridium difficile*-associated diarrhea in patient with cancer treated with fidaxomicin or vancomycin. J Clin Oncol. 2013;31:2493–9.
102. Cammarota G, Masucci L, Ianiro G, et al. Randomised clinical trial: faecal microbiota transplantation by colonoscopy vs. vancomycin for the treatment of recurrent *Clostridium difficile infection*. Ailment Pharmacol Ther. 2015;41:835–43.
103. Kelly CR, Ihunnah C, Fischer M, et al. Fecal microbiota transplant for treatment of *Clostridium difficile* infection in immunocompromised patients. Am J Gastroenterol. 2014;109:1065–71.
104. Webb BJ, Brunner A, Ford CD, et al. Fecal microbiota transplantation (FMT) for recurrent *Clostridium difficile* infection in hematopoietic stem cell transplant recipients. Transpl Infect Dis. 2016;18(4):628–33.
105. Gerding DN, Kelly CP, Rahav G, et al. Bezlotoxumab for prevention of recurrent Clostridium difficile infection in patients at increased risk for recurrence. Clin Infect Dis. 2018;67(5):649–56.
106. Mehta A, Rangarajan S, Borate U. A cautionary tale for probiotic use in hematopoietic SCT patients-Lactobacillus acidophilus sepsis in a patient with mantle cell lymphoma undergoing hematopoietic SCT. Bone Marrow Transplant. 2013;48:461–2.
107. Enache-Angoulvant A, Hennequin C. Invasive Saccharomyces infection: a comprehensive review. Clin Infect Dis. 2005;41:1559–68.
108. Hennequin C, Kauffmann-Lacroix C, Jobert A, et al. Possible role of catheters in *Saccharomyces boulardii* fungemia. Eur J Clin Microbiol Infect Dis. 2000;19:16–20.
109. Gouriet F, Million M, Henri M, Fournier PE, Raoult D. *Lactobacillus rhamnosus* bacteremia: an emerging clinical entity. Eur J Clin Microbiol Infect Dis. 2012;31:2469–80.
110. Carigan A, Poulin S, Martin P, et al. Efficacy of secondary prophylaxis with vancomycin for preventing recurrent Clostridium difficile infection in patients treated with systemic antimicrobial agents. Am J Gastroenterol. 2016;111(12):1834–40.
111. Van Hise NW, Cryant AM, Hennessey EK, et al. Efficacy of oral vancomycin in preventing recurrent Clostridium difficile infection in patients treated with systemic antimicrobial agents. Clin Infect Dis. 2016;63(5):651–3.
112. Mullane KM, Winston DJ, Nooka A, et al. A randomized, placebo-controlled trial of fidaxomicin for prophylaxis of clostridium difficile-associated diarrhea in adults undergoing hematopoietic stem cell transplantation. Clin Infect Dis. 2019;68(2):196–203.
113. UNAIDS. Global HIV & AIDS statistics – fact sheet; 2021. http://www.unaids.org/en/resources/fact-sheet. Accessed 24 July 2022.
114. Centers for Disease Control and Prevention. HIV in United States and dependent areas; 2020. https://www.cdc.gov/hiv/statistics/overview/ataglance.html. Accessed 24 July 2022.
115. Monitoring selected national HIV prevention and care objectives by using HIV surveillance data—United States and 6 dependent areas, 2019. HIV Surveillance Supplemental Report 2021;26(2). Accessed 24 July 2022.
116. Berry SA, Fleishman JA, Moore RD, Gebo KA, HIV Research Network. Trends in reasons for hospitalization in a multisite United States cohort of persons living with HIV, 2001-2008. J Acquir Immune Defic Syndr. 2012;59(4):368–75.
117. Davy-Mendez T, Napravnik S, Hogan BC, et al. Hospitalization rates and causes among persons with HIV in the United States and Canada, 2005–2015. J Infect Dis. 2021;223(12):2113–23.
118. Murray C, Ikuta KS, Sharara F, et al. Global burden of bacterial antimicrobial resistance in 2019: a systematic analysis. Lancet. 2022;399(10325):629–55.
119. Teeraananchai S, Kerr SJ, Amin J, Ruxrungtham K, Law MG. Life expectancy of HIV-positive people after starting combination antiretroviral therapy: a meta-analysis. HIV Med. 2017;18:256–66.
120. Summary of the HIV/AIDS Epidemic, 2020. UNAIDS/WHO estimates. WHO. https://www.who.int/data/gho/data/themes/hiv-aids. Accessed 24 July 2020.
121. Olaru I, Taconelli E, Yeung S, et al. The association between antimicrobial resistance and HIV infection: a systematic review and meta-analysis. Clin Microbiol Infect. 2021;27:846–53.

122. CDC. Antibiotic resistance threats in the United States, 2019. Atlanta: U.S. Department of Health and Human Services, CDC; 2019. https://www.cdc.gov/drugresistance/pdf/threats-report/2019-ar-threats-report-508.pdf. Accessed 25 Aug 2022.
123. Fishman N. Policy statement on antimicrobial stewardship by the Society for Healthcare Epidemiology of America (SHEA), the Infectious Diseases Society of America (IDSA), and the Pediatric Diseases Society (PIDS). Infect Control Hosp Epidemiol. 2012;33:322–7.
124. World Health Organization. Global tuberculosis report 2020. Available from: https://www.who.int/tb/publications/global_report/en/. Accessed 11 Aug 2022.
125. UNAIDS. Tuberculosis. https://www.unaids.org/en/topic/tuberculosis. Accessed 23 Aug 2022.
126. Joseph J, Rodvold KA. The role of carbapenems in the treatment of severe nosocomial respiratory tract infections. Expert Opin Pharmacother. 2008;9(4):561–75.
127. Treatment Action Group. Public comment on identifying root causes drug shortages and finding enduring solutions; 2019. http://www.treatmentactiongroup.org/content/public-comment-identifying-root-causes-drug-shortages-and-finding-enduring-solutions. Accessed 13 Feb 2019.
128. Schellack N, Benjamin D, Brink A, et al. A situational analysis of current antimicrobial governance, regulation, and utilization in South Africa. Int J Infect Dis. 2017;64:100–6.
129. Singh S, Menon V, Pp B, et al. Impact of a novel multidisciplinary anti-tubercular stewardship program in a tertiary care center in India. Open Forum Infect Dis. 2018;5(Suppl 1):S505.
130. Prabhu BP, Kunoor A, Sudhir S, et al. Anti-tuberculosis treatment stewardship in a private tertiary care hospital in South India. Public Health Action. 2018;8(3):151.
131. Padayatchi N, Mahomed S, Loveday M, Naidoo K. Antibiotic stewardship for drug resistant tuberculosis. Expert Opin Pharmacother. 2016;17(15):1981–3.
132. Carr W, Kurbatova E, Starks A, Goswami N, Allen L, Winston C. Interim guidance: 4-month rifapentine-moxifloxacin regimen for the treatment of drug-susceptible pulmonary tuberculosis—United States, 2022. MMWR Morb Mortal Wkly Rep. 2022;71:285–9. https://doi.org/10.15585/mmwr.mm7108a1.
133. Migliori GB, Langendam MW, D'Ambrosio L, et al. Protecting the tuberculosis drug pipeline: stating the case for the rational use of fluoroquinolones. Eur Respir J. 2012;40:814–22.
134. Jabeen K, Shakoor S, Hasan R. Fluoroquinolone-resistant tuberculosis: implications in settings with weak healthcare systems. Int J Infect Dis. 2015;3232:118–23.
135. Centers for Disease Control and Prevention. Drug-resistant TB. https://www.cdc.gov/tb/topic/drtb/default.htm#:~:text=Sometimes%20drug%2Dresistant%20TB%20occurs,from%20one%20person%20to%20another. Accessed 23 Aug 2022.
136. Deutsch-Feldman M, Pratt RH, Price SF, Tsang CA, Self JL. Tuberculosis—United States, 2020. MMWR Morb Mortal Wkly Rep. 2021;70:409–14.
137. Reported Tuberculosis in the United States, 2020. Drug-Resistant TB. Division of Tuberculosis Elimination, National Center for HIV, Viral Hepatitis, STD, and TB Prevention, Centers for Disease Control and Prevention. https://www.cdc.gov/tb/statistics/reports/2020/drug_resistant.htm. Accessed 11 Aug 2022.
138. World Health Organization. Tuberculosis 2021. https://www.who.int/news-room/fact-sheets/detail/tuberculosis#:~:text=Multidrug%2Dresistant%20TB%20(MDR%2D,the%20cumulative%20reduction%20was%2011%25. Accessed 23 Aug 2022.
139. World Health Organization. The end TB strategy. 2014. https://www.who.int/tb/strategy/en/. Accessed 13 Feb 2019.
140. Treatment Action Group. Pipeline report 2021 tuberculosis treatment; 2021. https://www.treatmentactiongroup.org/wp-content/uploads/2021/11/pipeline_TB_Treatment_2021_final.pdf. Accessed 11 Aug 2022.
141. Boot M, Commandeur S, Subudhi AK, et al. Accelerating early antituberculosis drug discovery by creating mycobacterial indicator strains that predict mode of action. Antimicrob Agents Chemother. 2018;62(7):e00083–18.
142. Katzir I, Cokol M, Aldridge BB, Alon U. Prediction of ultra-high-order antibiotic combinations based on pairwise interactions. PLoS Comput Biol. 2019;15(1):e1006774.

143. Centers for Disease Control and Prevention. Estimated HIV incidence and prevalence in the United States, 2015–2019. HIV Surveil Suppl Rep. 2021;26(1) http://www.cdc.gov/hiv/library/reports/hiv-surveillance.html. Published May 2021. Accessed 11 Aug 2022.
144. Centers for Disease Control and Prevention. Sexually transmitted disease surveillance 2020. Gonorrhea. Atlanta: US Department of Health and Human Services. https://www.cdc.gov/std/statistics/2020/overview.htm#Gonorrhea. Accessed 23 Aug 2022.
145. STD surveillance 2020. Division of STD Prevention, National Center for HIV, Viral Hepatitis, STD, and TB Prevention, Centers for Disease Control and Prevention. https://www.cdc.gov/std/statistics/2020/default.htm. Accessed 11 Aug 2022.
146. Antibiotic-Resistant Gonorrhea Basic Information. Division of STD Prevention, National Center for HIV, Viral Hepatitis, STD, and TB Prevention, Centers for Disease Control and Prevention https://www.cdc.gov/std/gonorrhea/arg/basic.htm. Accessed 11 Aug 2022.
147. Workowski KA, Bachmann LH, Chan PA, et al. Sexually transmitted infections treatment guidelines, 2021. MMWR Recomm Rep. 2021;70(4):1–187.
148. Gonococcal Isolate Surveillance Project (GISP) Division of STD Prevention, National Center for HIV, Viral Hepatitis, STD, and TB Prevention, Centers for Disease Control and Prevention. https://www.cdc.gov/std/gisp/default.htm. Accessed 11 Aug 2022.
149. Centers for Disease Control and Prevention. Sexually Transmitted diseases (STDs). STD Surveillance Network (SSuN) https://www.cdc.gov/std/ssun/default.htm. Accessed 11 Aug 2022.
150. Ellis O, Hemarajata P, Shahkolahi A, et al. A multisite implementation of a real-time polymerase chain reaction assay to predict ciprofloxacin susceptibility in Neisseria gonorrhoeae. Diagn Microbiol Infect Dis. 2019;94(3):213–7.
151. Chen L, Shin DJ, Zheng S, Melendez JH, Gaydos CA, Wang TH. Direct-qPCR assay for coupled identification and antimicrobial susceptibility testing of *Neisseria gonorrhoeae*. ACS Infect Dis. 2018;4(9):1377–84.
152. Abrams AJ, Trees DL. Genomic sequencing of *Neisseria gonorrhoeae* to respond to the urgent threat of antimicrobial-resistant gonorrhea. Pathog Dis. 2017;75(4):1–9.
153. Gardy JL, Johnston JC, Ho Sui SJ, et al. Whole-genome sequencing and social-network analysis of a tuberculosis outbreak. N Engl J Med. 2011;364:730–9.
154. Alirol E, Wi TE, Bala M, et al. Multidrug-resistant gonorrhea: a research and development roadmap to discover new medicines. PLoS Med. 2017;14(7):e1002366.
155. Sevilla JP, Bloom DE, Cadarette D, Jit M, Lipsitch M. Toward economic evaluation of the value of vaccines and other health technologies in addressing AMR. Proc Natl Acad Sci U S A. 2018;115(51):12911–9.
156. Abara WE, Bernstein KT, Lewis FM, et al. Effectiveness of a serogroup B outer membrane vesicle meningococcal vaccine against gonorrhea: a retrospective observational study. Lancet Infect Dis. 2022;22(7):P1021–9.
157. Chen SX, Yin YP, Li XY. A ROADMAP plan to address research needs for gonococcal antimicrobial resistance in China. Clin Infect Dis. 2019;68(3):505–10.
158. World Health Organization. Salmonella; 2018. https://www.who.int/news-room/fact-sheets/detail/typhoid. Accessed 28 Mar 2019.
159. Centers for Disease Control and Prevention, National Center for Emerging and Zoonotic Infectious Diseases (NCEZID), Division of Foodborne, Waterborne, and Environmental Diseases (DFWED) Salmonella. https://www.cdc.gov/salmonella/index.html. Accessed 11 Aug 2022.
160. Reddy EA, Shaw AV, Crump JA. Community-acquired bloodstream infections in Africa: a systematic review and meta-analysis. Lancet Infect Dis. 2010;10:417–32.
161. Scallan E, Hoekstra RM, Angulo FJ, et al. Foodborne illness acquired in the United States – major pathogens. Emerg Infect Dis. 2011;17(1):7–15.
162. CDC. Antibiotic resistance threats in the United States, 2013. Atlanta: U.S. Department of Health and Human Services, CDC; 2013. https://www.cdc.gov/drugresistance/pdf/ar-threats-2013-508.pdf. Accessed 12 Aug 2022.

163. Centers for Disease Control and Prevention, National Center for Emerging and Zoonotic Infectious Diseases (NCEZID), Division of Foodborne, Waterborne, and Environmental Diseases (DFWED) Salmonella. https://www.cdc.gov/salmonella/general/technical.html. Accessed 12 Aug 2022.
164. Haines C, Sulkowski M. Chapter 126: Gastrointestinal, hepatobiliary, and pancreatic manifestations of human immunodeficiency virus infection. In: Mandell, Douglas, and Bennett's principles and practice of infectious diseases, 9th ed.; 2014. p. 1567–1573.
165. Sanchez TH, Brooks JT, Sullivan PS, et al. Bacterial diarrhea in persons with HIV infection, United States, 1992-2002. Clin Infect Dis. 2005;41(11):1621–7.
166. Center for Disease Control and Prevention. National Antimicrobial Resistance Monitoring System (NARMS); 2017. https://wwwn.cdc.gov/narmsnow/. Accessed 15 Mar 2019.
167. Chatham-Stephens K, Medalla F, Hughes M, et al. Emergence of extensively drug-resistant salmonella typhi infections among travelers to or from Pakistan—United States, 2016–2018. MMWR Morb Mortal Wkly Rep. 2019;68:11–3. https://doi.org/10.15585/mmwr.mm6801a3externalicon.
168. World Health Organization. Typhoid fact sheet; 2018. https://www.who.int/mediacentre/factsheets/typhoid/en/. Accessed 28 Mar 2019.
169. Park SE, Pham DT, Boinett C, et al. The phylogeography and incidence of multi-drug resistant typhoid fever in sub-Saharan Africa. Nat Commun. 2018;9(1):5094.
170. Tacconelli E, Carrara E, Savoldi A, et al. Discovery, research, and development of new antibiotics: the WHO priority list of antibiotic-resistant bacteria and tuberculosis. Lancet Infect Dis. 2018;18(3):318–27.
171. Obi CL, Onabolu B, Momba MNB, et al. The interesting cross-paths of HIV/AIDS and water in Southern Africa with special reference to South Africa. Water SA. 2006;32(3):323–43.
172. United Nations. Sustainable development goals: clean water and sanitation. https://www.un.org/sustainabledevelopment/water-and-sanitation/. Accessed 28 Mar 2019.
173. United States Department of Health and Human Services. Food safety for HIV/AIDS patients. https://www.foodsafety.gov/risk/hiv/index.html. Accessed 28 Mar 2019.
174. United States Energy Information Administration. The effect of income on appliances in U.S. households. https://www.eia.gov/consumption/residential/data/2001/appliances/appliances.php. Accessed 28 Mar 2019.
175. United States Energy Information Administration. Global access to electricity has increased the last two decades; 2017. https://www.eia.gov/todayinenergy/detail.php?id=31552. Accessed 28 Mar 2019.
176. Center for Disease Control and Prevention. Saving lives by taking a One Health Approach; 2017. https://www.cdc.gov/onehealth/multimedia/factsheet.html. Accessed 28 Mar 2019.
177. Panel on Opportunistic Infections in HIV-Infected Adults and Adolescents. Guidelines for the prevention and treatment of opportunistic infections in HIV-infected adults and adolescents: recommendations from the Centers for Disease Control and Prevention, the National Institutes of Health, and the HIV Medicine Association of the Infectious Diseases Society of America. http://aidsinfo.nih.gov/contentfiles/lvguidelines/adult_oi.pdf. Accessed 28 Mar 2019.
178. CDC. National Center for Immunization and Respiratory Diseases. Pneumococcal vaccination: what everyone should know. https://www.cdc.gov/vaccines/vpd/pneumo/public/index.html#:~:text=Prevnar%2020%C2%AE%20%3A%20Doctors%20give,cause%20serious%20infections%20in%20adults. Accessed 12 Aug 2022.
179. Kim L, McGee S, Tomczyk S, et al. Biological and epidemiological features of antibiotic-resistant *Streptococcus pneumoniae* in pre- and post-conjugate vaccine eras: a United States perspective. Clin Microbiol Rev. 2016;29(3):525–55.
180. Chiu CH, Su LH, Huang YC, et al. Increasing ceftriaxone resistance and multiple alterations of penicillin-binding proteins among penicillin-resistant *Streptococcus pneumoniae* isolates in Taiwan. Antimicrob Agents Chemother. 2007;51(9):3404–6.

181. Jordano Q, Falcó V, Almirante B, et al. Invasive pneumococcal disease in patients infected with HIV: still a threat in the era of highly active antiretroviral therapy. Clin Infect Dis. 2004;38(11):1623–8.
182. Meynard JL, Barbut F, Blum L, et al. Risk factors for isolation of *Streptococcus pneumoniae* with decreased susceptibility to penicillin G from patients infected with human immunodeficiency virus. Clin Infect Dis. 1996;22(3):437–40.
183. Hamel MJ, Greene C, Chiller T, et al. Does cotrimoxazole prophylaxis for the prevention of HIV-associated opportunistic infections select for resistant pathogens in Kenyan adults? Am J Trop Med Hyg. 2008;79(3):320–30.
184. Powis KM, Souda S, Lockman S, et al. Cotrimoxazole prophylaxis was associated with enteric commensal bacterial resistance among HIV-exposed infants in a randomized controlled trial, Botswana. J Int AIDS Soc. 2017;20(3):e25021.
185. World Health Organization. Antibiotic resistance: why vaccination is important. https://www.who.int/news-room/questions-and-answers/item/antibiotic-resistance-why-vaccination-is-important. Accessed March 2019.
186. Billedo JA, et al. Evaluating the impact of a pharmacist-led antiretroviral stewardship program on reducing drug interactions in HIV-infected patients. J Int Assoc Provid AIDS Care. 2016;15(1):84–8.
187. Hsu AJ, et al. Antiretroviral stewardship in a pediatric HIV clinic: development, implementation and improved clinical outcomes. Pediatr Infect Dis J. 2016;35(6):642–8.
188. Sanders J, et al. Antimicrobial stewardship program to reduce antiretroviral medication errors in hospitalized patients with human immunodeficiency virus infection. Infect Control Hosp Epidemiol. 2014;35(3):272–7.
189. Koren DE, Scarsi KK, Farmer EK, et al. A call to action: the role of antiretroviral stewardship in inpatient practice, a joint policy paper of the Infectious Diseases Society of America. Clin Infect Dis. 2020;70(11):2241–6.
190. Ma A, Chen DM, Chau FM, Saberi P. Improving adherence and clinical outcomes through an HIV pharmacist's interventions. AIDS Care. 2010;22(10):1189–94.
191. Li EH, Foisy MM. Antiretroviral and medication errors in hospitalized HIV-positive patients. Ann Pharmacother. 2014;48(8):998–1010.
192. Evans-Jones JG, Cottle LE, Back DJ, et al. Recognition of risk for clinically significant drug interactions among HIV-infected patients receiving antiretroviral therapy. Clin Infect Dis. 2010;50(10):1419–21.
193. Hirsch J, Gonzales A, Rosenquist T, et al. Antiretroviral therapy adherence, medication use, and health care costs during 3 years of a community pharmacy medication therapy management program for medical beneficiaries with HIV/AIDS. J Manag Care Spec Pharm. 2011;17(3):213–23.
194. O'Connor J, Viecha MJ, Phillips AN, et al. Effect of immediate initiation of antiretroviral therapy on risk of severe bacterial infections in HIV-positive people with CD4 cell counts of more than 500 cells per μL: secondary outcome results from a randomised controlled trial. Lancet. 2017;4(3):e105–12.
195. Halperin J, Conner K, Butler I, et al. A care continuum of immediate ART for newly diagnosed patients & patients presenting later to care at a federally qualified health center in New Orleans. Open Forum Infect Dis. 2019;6:ofz161.
196. Halperin J, Butler I, Conner K, et al. Linkage and antiretroviral therapy within 72 hours at a federally qualified health center in New Orleans. AIDS Patient Care STDs. 2018;32(2):39–41.
197. Centers for Medicare and Medicaid Services (CMS). Omnibus burden reduction (conditions of participation). Final rule CMS-3346-F; 2019. https://www.cms.gov/newsroom/fact--sheets/omnibus-burden-reduction-conditions-participation-final-rule-cms-3346-f. Accessed 5 June 2022.
198. Seo SK, Lo K, Abbo LM. Current state of antimicrobial stewardship at solid organ and hematopoietic cell transplant centers in the US. Infect Control Hosp Epidemiol. 2016;37(10):1195–200.

199. Ziakas PD, Pliakos EE, Zervou FN, Knoll BM, Rice LB, Mylonakis E. MRSA and VRE colonization in solid organ transplantation: a meta-analysis of published studies. Am J Transplant. 2014;14(8):1887–94.
200. Bert F, et al. Risk factors for Staphylococcus aureus infection in liver transplant recipients. Liver Transpl. 2005;11(9):1093–9.
201. Singh N, et al. Methicillin-resistant Staphylococcus aureus: the other emerging resistant gram-positive coccus among liver transplant recipients. Clin Infect Dis. 2000;30(2):322–7.
202. Shields RK, et al. Staphylococcus aureus infections in the early period after lung transplantation: epidemiology, risk factors, and outcomes. J Heart Lung Transplant. 2012;31(11):1199–206.
203. Giannella M, et al. Carbapenemase-producing Enterobacteriaceae in transplant patients. J Antimicrob Chemother. 2021;76(Supplement_1):i27–39.
204. Freire MP, et al. Carbapenem-resistant Enterobacteriaceae acquired before liver transplantation: impact on recipient outcomes. Transplantation. 2017;101(4):811.
205. Pouch SM, Satlin MJ. Carbapenem-resistant Enterobacteriaceae in special populations: solid organ transplant recipients, stem cell transplant recipients, and patients with hematologic malignancies. Virulence. 2017;8(4):391–402. https://doi.org/10.1080/21505594.2016.1213472.
206. Vallés J, Rello J, Ochagavía A, Garnacho J, Alcalá MA. Community-acquired bloodstream infection in critically ill adult patients: impact of shock and inappropriate antibiotic therapy on survival. Chest. 2003 May;123(5):1615–24. https://doi.org/10.1378/chest.123.5.1615.
207. Averbuch D, et al. Antimicrobial resistance in gram-negative rods causing bacteremia in hematopoietic stem cell transplant recipients: intercontinental prospective study of the Infectious Diseases Working Party of the European Bone Marrow Transplantation Group. Clin Infect Dis. 2017;65(11):1819–28.
208. Harbarth S, Garbino J, Pugin J, Romand JA, Lew D, Pittet D. Inappropriate initial antimicrobial therapy and its effect on survival in a clinical trial of immunomodulating therapy for severe sepsis. Am J Med. 2003;115(7):529–35. https://doi.org/10.1016/j.amjmed.2003.07.005.
209. Ibrahim EH, Sherman G, Ward S, Fraser VJ, Kollef MH. The influence of inadequate antimicrobial treatment of bloodstream infections on patient outcomes in the ICU setting. Chest. 2000 Jul;118(1):146–55. https://doi.org/10.1378/chest.118.1.146.
210. Pillinger KE, et al. Inpatient antibiotic stewardship interventions in the adult oncology and hematopoietic stem cell transplant population: a review of the literature. Ann Pharmacother. 2020;54(6):594–610.
211. Ilges D, et al. Assessment of antibiotic de-escalation by spectrum score in patients with nosocomial pneumonia: a single-center, retrospective cohort study. In: Open forum infectious diseases, Vol. 8, No. 11. Oxford University Press; 2021.
212. Green M, Blumberg EA, Danziger-Isakov L, Huprikar S, Kotton CN, Kumar D. Foreword: 4th edition of the American society of transplantation infectious diseases guidelines. Clin Transpl. 2019;33(9):e13642.
213. Ouyang D, et al. Incidence of temporary mechanical circulatory support before heart transplantation and impact on post-transplant outcomes. J Heart Lung Transplant. 2018;37(9):1060–6.
214. Molina EJ, et al. The society of thoracic surgeons intermacs 2020 annual report. Ann Thorac Surg. 2021;111(3):778–92.
215. Shah P, et al. Stroke and death risk in ventricular assist device patients varies by ISHLT infection category: an INTERMACS analysis. J Heart Lung Transplant. 2019;38(7):721–30.
216. Farr MA, Stehlik J. Heart xenotransplant: a door that is finally opening. Circulation. 2022;145:871–3.
217. Fishman JA. Infectious disease risks in xenotransplantation. Am J Transplant. 2018;18(8):1857–64.
218. Centers for Disease Control and Prevention. Core elements of antibiotic stewardship. Atlanta: U.S. Department of Health & Human Services; 2022. [Available from: https://www.cdc.gov/antibiotic-use/core-elements/index.html. Accessed 3 Jan 2022. Practical guidance on the implementation of antimicrobial stewardship programs in hospital setting]

219. So M, Walti L. Challenges of antimicrobial resistance and stewardship in solid organ transplant patients. Curr Infect Dis Rep. 2022;24:1–13.
220. Snyder M, Pasikhova Y, Baluch A. Early antimicrobial de-escalation and stewardship in adult hematopoietic stem cell transplantation recipients: retrospective review. In: Open forum infectious diseases, vol. 4, No. 4. Oxford University Press: US; 2017.
221. So M, et al. Antimicrobial stewardship by academic detailing improves antimicrobial prescribing in solid organ transplant patients. Eur J Clin Microbiol Infect Dis. 2019;38(10):1915–23.
222. Rosa R, Simkins J, Camargo JF, Martinez O, Abbo LM. Solid organ transplant antibiograms: an opportunity for antimicrobial stewardship. Diagn Microbiol Infect Dis. 2016;86(4):460–3.
223. Hamandi B, et al. Impact of infectious disease consultation on the clinical and economic outcomes of solid organ transplant recipients admitted for infectious complications. Clin Infect Dis. 2014;59(8):1074–82.
224. Micallef C, et al. Introduction of an antifungal stewardship programme targeting high-cost antifungals at a tertiary hospital in Cambridge, England. J Antimicrob Chemother. 2015;70(6):1908–11.

Antimicrobial Stewardship in the Intensive Care Unit

8

Iris H. Chen, David P. Nicolau, and Joseph L. Kuti

Introduction

Demand for critical care services in the USA has grown tremendously over the years [1]. From 2002 to 2009, intensive care units (ICUs) in the USA saw nearly a 50% increase in admissions from their emergency departments [2]. While patients are admitted to the ICU for a variety of reasons, the composition has shifted over the years. A study on Medicare beneficiaries noted a decrease in admissions for cardiac diseases and an increase for infections; this was accompanied with increased rates of acute organ dysfunction and invasive medical procedures [3]. Simply put, today's ICU patients are sicker and infections are frequently implicated. Healthcare-associated infections (HAI) generally occur in approximately 4–6% of hospital patients, but ICU patients are roughly three times more likely to be affected [4, 5]. Common HAIs include pneumonia, bloodstream infection, surgical-site infection, as well as device-associated infections, including those of the urinary tract. Ventilator-associated pneumonia (VAP), in particular, affects up to one quarter of intubated patients, making it a leading cause for ICU antibiotic use [6–9]. In 2010, sepsis was the most frequently assigned discharge diagnosis for patients in the ICU [3]. Depending on severity, sepsis and septic shock can result in up to 80% mortality [10].

Given the frequency of infection, whether acquired from the community or a healthcare environment, it is unsurprising that antibiotics compose a sizable fraction of dispensed medications in the ICU setting. The Extended Prevalence of Infection in Intensive Care study found that 71% of ICU patients globally were receiving antibiotics on the day of the study [11]. Unfortunately, many of these antibiotic prescriptions might be considered inappropriate. Not unlike other practice settings,

I. H. Chen · D. P. Nicolau · J. L. Kuti (✉)
Center for Anti-Infective Research and Development, Hartford Hospital, Hartford, CT, USA
e-mail: David.Nicolau@hhchealth.org; Joseph.Kuti@hhchealth.org

S. Doron, M. Campion (eds.), *Antimicrobial Stewardship in Non-Traditional Settings*, https://doi.org/10.1007/978-3-031-21710-4_8

inappropriate antibiotic therapy in the ICU includes administration of broad-spectrum antibiotic(s), including combination therapy, for longer than clinically indicated (i.e., lack of de-escalation, discontinuation, or ignoring standard treatment durations) and empirically starting too narrow-spectrum therapy due to under appreciation for local resistance rates [12, 13]. Additionally, due to substantial acute and chronic changes in organ function, pharmacokinetics can be altered in critically ill patients leading to over- or under-dosing, thereby resulting in adverse events or development of antibiotic resistance [14]. Lastly, novel laboratory diagnostics are frequently targeted at ICU patients and the proper selection and interpretation of these tests can further complicate the administration or management of appropriate antibiotic therapy [15, 16].

Antibiotic resistance is often considered more frequent in the ICU than on general wards, but this is not always the case [17]. True ICU resistance is often unknown as most hospitals use a single antibiogram institution wide, thus preventing local providers from understanding if differences in resistance are apparent between ICUs and general wards in their hospital [18]. The perception of greater resistance combined with the acute severity of illness of patients in the ICU can lead to empiric administration of broad-spectrum antibiotic therapy regardless of actual prevalence or consideration of the differential of causative pathogens for certain infections.

As a result, antimicrobial stewardship programs (ASPs) have become a necessity in today's ICU. Defined as "coordinated interventions designed to improve and measure the appropriate use of [antibiotic] agents by promoting the selection of the optimal drug regimen including dosing, duration of therapy, and route of administration," antimicrobial stewardship can result in benefits at multiple levels [19]. While the Joint Commission has recognized the benefits of ASPs and recently made it a requirement for acute care hospitals to establish a program, no specific guidance on metrics or implementation standards has been publicized yet. General standards include proof of ASP development and implementation, staff and patient education, access to infectious disease (ID) specialists where available, and documentation of ASP protocols and the efforts to improve them. This chapter will review the specific ASP strategies likely to have the greatest impact in the ICU setting.

Components of an ICU Antimicrobial Stewardship Program

ASP Intervention(s)

There are two main interventions utilized by contemporary ASPs: pre-prescription authorization (i.e., antibiotic restriction) and post-prescription audit and feedback, as well as various supplemental strategies to aid in achieving ASP goals of reducing inappropriate antimicrobial use and reducing antimicrobial resistance (Table 8.1). Prior to the original ASP guidelines published by the Infectious Diseases Society of America (IDSA), most hospitals used antibiotic restriction as the primary intervention to manage (or prevent) inappropriate antibiotic use [20]. A number of studies successfully demonstrate the ability for restriction programs to reduce antibiotic

Table 8.1 Interventions in an ICU Antimicrobial Stewardship Program

Intervention	Pros	Cons
Pre-prescription authorization (antibiotic restriction)	– Effectively reduces defined daily doses (DDD) or days of therapy (DOT) for target antibiotics – Decreases resistance of restricted drug – Low time commitment	– Potential increases in use of non-restricted antibiotics and resistance rates – Does not provide opportunity for addressing utilization of unrestricted drugs or DOT after starting antibiotics
Post-prescription audit and feedback	– Patient-specific recommendations – Allows assessments of de-escalation and duration of therapy – Education opportunities to improve future antibiotic utilization – Effectively reduces DDD or DOT for individual patients – No harmful effects on patient outcomes	– Substantial time commitment required – Outcomes dependent upon acceptances of recommendations
Indication-specific guidelines	– Evidence-based and customizable based on institutional epidemiology and type of infection – Increases rates of appropriate empiric antibiotic therapy – Reduces mortality associated with inappropriate empiric antibiotic therapy	– Requires periodic updates – No guarantee that providers will use guideline so ongoing education and promotion required
ICU-specific antibiogram	– Accounts for unit pathogen prevalence and local resistance to more accurately guide empiric therapy in the ICU	– Requires annual updates – Reported susceptibility rates inaccurate for low frequency pathogens
Antibiotic cycling/mixing	– Heterogeneous antibiotic use may reduce and prevent antibiotic resistance	– Little data to support that intervention works or is sustainable – May not account for local resistance outbreaks in unit or previous antibiotic therapy – Difficult to operationalize

defined daily doses (DDD) or days of therapy (DOT) for targeted agents in the ICU setting [21–24]. Decreases in use were often accompanied by reductions in resistance to these antibiotics. Not all restriction studies, however, have observed improvements in resistance rates with accompanying DDD or DOT reductions [25]. Without additional intervention and education, other "non-restricted" antibiotics simply replace those agents, endangering these antibiotics to the development of resistance. In one of the earliest and most publicized studies of antibiotic restriction, a large hospital in Queens NY restricted all cephalosporin antibiotics in response to the increasing prevalence of extended spectrum β-lactamase (ESBL) producing *Klebsiella pneumoniae* [26]. Although this restriction was accompanied by a 70.9% reduction in the incidence of ceftazidime-resistant *Klebsiella* infection in the ICU ($p < 0.001$), empiric antibiotic prescriptions were replaced with imipenem resulting in a 68.7% increase in imipenem-resistant *Pseudomonas aeruginosa* throughout the medical center. Importantly, a pre-prescription authorization only approach misses critical interventions that occur during antimicrobial therapy such as opportunities for de-escalation and discontinuation of antibiotics.

More recently, post-prescription audit and feedback has replaced or been added to pre-prescription authorization to improve ASP interactions and outcomes [27, 28]. This intervention permits opportunity for communication between the ASP team and the clinical team to provide patient-specific recommendations as well as education with the intent the clinical team would apply the learned strategy to future similar scenarios. It also addresses antimicrobial appropriateness at all times during therapy. The greatest challenge to post-prescription audit and feedback remains the time commitment to physically conduct the review and provide education, as well as the requirement that clinical teams, in the end, accept the ASP recommendation. In light of this challenge, this approach may be even more preferable in the ICU setting where there are lower bed counts, and the patient review is more focused, albeit more complicated due to severity of illness.

To address which strategy is more effective, Tamma and colleagues conducted a quasi-experimental, crossover time series analysis in their hospital [29]. For the first 4 months, 2 medicine teams were assigned to the pre-prescription authorization arm and the other 2 teams to post-prescription audit and feedback. The teams were then assigned to the alternative regimen for the final 4 months of the study. The authors found that post-prescription audit and feedback was more effective at decreasing antibiotic DOTs; however, no differences in clinical outcomes (e.g., *Clostridioides difficile* infection, hospital length of stay, or mortality) were observed. Importantly, ICUs were excluded from their study, and the clinical implications (i.e., could such programs worsen clinical outcomes in more severely ill patients) are difficult to extrapolate.

A recent systematic review of post-prescription audit and feedback interventions conducted only in the ICU observed that these programs can indeed be successfully implemented in this setting without adversely affecting mortality [30], which is reassuring and essential for critical care providers to appreciate. The pooled relative risk for ICU mortality associated with audit and feedback in the included studies was 1.03 (96% confidence interval [CI], 0.93–1.14). In the largest and longest duration study of its kind, Morris and colleagues implemented an antimicrobial stewardship program in 4 academic ICUs between 2007–2015 [31]. The stewardship intervention consisted solely of post-prescription audit and feedback (which the authors cleverly termed "coaching") at least 3 times weekly with the ICU team. These 15-minute coaching meetings were performed by the stewardship pharmacist and included review of all antibiotic regimens in all patients on the unit at the time of the review. The study intervention observed significant reductions in DDD/100 patient-days across all antibiotic classes without any significant changes in mortality, length of stay, or readmission, again suggesting a reduction in antibiotic utilization can be accomplished without any ill effects on clinical endpoints. Annual *P. aeruginosa* and *Escherichia coli* susceptibility rates for meropenem, ceftazidime (ceftriaxone for *E. coli*), tobramycin (gentamicin for *E. coli*), and ciprofloxacin remained unchanged or improved after implementation of the stewardship interventions in each ICU.

Given the above examples, a combination of both interventions, restriction of certain antimicrobials and post-prescription audit and feedback for others, should lead to the least amount of inappropriate antibiotic therapy in the ICU setting without adversely affecting clinical outcomes. A study in a French ICU over an 8-year period observed just that [32]. The ICU stewardship program combined pre-prescription authorization of specific antibiotics including carbapenem, fluoroquinolones, and third generation cephalosporins, with weekly post-prescription audit and feedback meetings with an ID specialist and microbiologist. Over the study period, 50–85% reductions in targeted antibiotics classes were observed; these reductions were replaced by increases in penicillins (with and without inhibitors), sulfonamides, and macrolides. Multivariate analyses demonstrated no negative impact on ICU mortality; furthermore, antibiotic resistance among *P. aeruginosa* improved across all drug classes.

The exact classes of antibiotics that require restriction may vary by ICU based on prevalence of local resistance phenotypes. There also may be clinical scenarios where an antibiotic class is restricted across the hospital, but acceptable for empiric use in specific situations. In the French multimodal stewardship study [32], carbapenems were restricted for empiric use unless a patient had septic shock plus history of ESBL in the previous six months, while fluoroquinolones were restricted to documented infections only, mainly osteoarticular infections. These situational restrictions cater to the severity of illness of ICU patients, the importance of "getting it right" up front, while taking into consideration local epidemiology. During a program aimed at improving outcomes in VAP patients across 3 ICUs, local epidemiology and phenotypic testing justified the empiric use of high-dose, prolonged infusion meropenem in one ICU to address the higher resistance rates among *P. aeruginosa* in that unit [33]. By using a computerized order entry system through a clinical pathway specific to VAP patients, meropenem was available to medical ICU providers without requiring pre-prescription authorization; meropenem was otherwise restricted throughout the rest of the institution. High-dose cefepime was the preferred empiric agent for VAP in the other ICUs secondary to local epidemiology in those units, and the computerized order entry system was designed to guide providers to the correct first-line choice by unit when VAP was suspected. Collectively, this targeted approach resulted in significant reductions in infection-related mortality, decreases in ICU length of stay and hospital costs, reductions in multidrug resistant superinfections, and no significant change in antibiotic cost to the pharmacy [33]. One notable experience during this program was a return to the selection of inappropriate empiric antibiotic therapy and a trend toward worse clinical outcomes when the stewardship team no longer audited and managed the use of the pathway, thereby further justifying the need for continuous post-prescription audit and feedback to ICU teams to maintain positive results [34].

Antibiotic cycling, or rotation, is an additional stewardship strategy historically applied in ICUs, where a specific antibiotic is preferentially used as first-line therapy in all patients requiring antibacterial therapy during a defined time period of time, after which another antibiotic becomes the preferred therapy and so on. In contrast to general hospital wards, this approach was popular in the ICU setting due to its secluded environment, focused patient populations and infection types, and fewer providers required to be in adherence with the protocol. The theoretical basis for antibiotic cycling is complex, based on controlled heterogeneity to reduce selective pressure and beyond the scope of this chapter. Neither version of the contemporary antibiotic stewardship guidelines recommends antibiotic cycling due to poor evidence that this reduction in selective pressure is achievable or sustainable [19, 20]. Notably, the majority of previous studies assessing antibiotic cycling yielded inconclusive results and suffered from various study design issues. More recently, the SATURN Consortium completed a large cluster-randomized crossover study in 8 European ICUs to address the methodological shortcomings of past studies [35]. ICUs were randomized to the empiric antibiotic selection using 6-week rotation cycles for each antibiotic class; the crossover intervention was an antibiotic mixing strategy, which has similar goals to rotation, but the empirical choice changes with each consecutive patient prescription. Overall, antibiotic use, resistance prevalence, and mortality were not different between interventions, and neither resulted in improvements over the baseline period. While the authors indicate the book has yet to be closed on antibiotic cycling (defining the optimum duration and rotation of antibiotics could yield improved results), the negative findings of this large study designed to address previous trial limitations make it challenging to recommend implementing an antibiotic cycling or mixing strategy in contemporary ICUs.

The use of clinical guidelines or pathways in the ICU is another stewardship approach to assist in guiding prescribers to the selection of the most appropriate antibiotic therapy. Given the majority of infections in the ICU setting are associated with the respiratory tract, urinary tract, or bloodstream, a focused pathway based on local epidemiology and resistance rates should be used to guide empiric antibiotic selection [36]. Ibrahim and colleagues implemented a clinical guideline to aid in antibiotic selection for the treatment of ventilator-associated pneumonia in their medical ICU [37]. The guideline pointed providers toward prescribing triple antibiotic coverage (vancomycin, imipenem, ciprofloxacin) based on local prevalence of pathogens and resistance rates. The guideline also recommended duration of therapy. The initial administration of appropriate antimicrobial treatment was statistically greater during the intervention period (94.2% versus 48.0%, $p < 0.001$), antibiotic duration was shorter (8.6 ± 5.1 days versus 14.8 ± 8.1, $p < 0.001$), and a second episode of ventilator-associated pneumonia occurred less often (7.7% versus 24.0%, $p = 0.030$). No difference in mortality or ICU length of stay was observed.

Personnel

Multidisciplinary personnel are required to develop and maintain an ASP in the ICU [38, 39]. Not unlike ASP leaders discussed in the other chapters of this book, it is our opinion that stewardship in the ICU is best led by an Infectious Diseases (ID) physician, ID pharmacist, or both, who have clinical expertise in ID and are trained in ASP strategies. One study found that ID physicians were able to aid stewardship efforts specifically by participating in the ICU's multidisciplinary rounds [40]. As noted by the author, the specialists provided their expertise in a number of ways that could have ultimately influenced the team's prescribing decisions. Such methods included expanding and refining the differential diagnosis, advising on diagnostic maneuvers, interpretation of infectious diseases-related tests, suggested empiric and/or specific and/or prophylactic antimicrobials, addressing adverse drug reaction issues, and providing advice on isolation procedures. The ICU saw a decrease in both DOTs and cost during days the ID physician had participated in rounds.

As seen in previously discussed studies, ID pharmacists are also optimal for carrying out stewardship responsibilities in the ICU [31, 33]. The Morris ICU ASP study [34] combined resources of an ID physician and ID pharmacist, but the authors noted that the ID pharmacists did the majority of coaching to the ICU team and embraced the physician's expertise on more challenging or contentious topics. The successful results identified by Morris and colleagues required only 1 h of preparation time and 15 min coaching sessions with the ICU team 3× weekly (although the program did start out with daily sessions). This reasonable time commitment permits the ASP team to continue to manage stewardship activities throughout the rest of the medical center.

Interestingly, there are little data to support ASP efforts in the ICU being led by intensivists or critical care pharmacists. That said, collaboration with these team members will be a vital component of a successful ICU stewardship intervention; furthermore, there are indeed ICUs where such practices have been very successful. The critical care team (intensivists, fellows, and critical care pharmacists) at a large academic US hospital has been quite successful in performing stewardship in their ICU for some time [37–39]. In a recent study, their team aimed at conducting enhanced antimicrobial de-escalation, essentially daily post-audit prescription audit and feedback, for mechanically ventilated pneumonia patients [41]. Compared with standard of care antibiotic management in their ICU, their intervention did not significantly reduce inappropriate antibiotic duration or increase de-escalation strategies because their critical care providers were already living these "best-practices." Notably, the ID trained ASP team at this hospital is also involved with guidance of antimicrobial practices in the ICU, importantly demonstrating that a collaborative, multidisciplinary team-based approach can be successful and sustainable.

No less important in the ICU ASP program is the clinical microbiology laboratory. Swift and accurate identification of ICU pathogens and the susceptibility data to go with it remain absolute necessities in effectively treating critically-ill infected patients. At the same time, it is important to be prudent with choosing which tests are routinely done and reported in order to avoid misinterpretation. On a broader scale, the microbiology laboratory also provides surveillance regarding institutional and local susceptibilities.

Antibiograms

Hospital antibiograms are a fundamental tool used by the microbiology department and ASP to identify annual (or more frequent) changes in antimicrobial susceptibility and highlight local epidemiology for the prescriber to use when making empiric antimicrobial decisions. The Clinical Laboratory Standards Institute (CLSI) provides guidance on analysis and presentation of cumulative antimicrobial susceptibility test data in their M39 guideline (www.CLSI.org). While most hospitals use a single antibiogram to report these data hospital-wide, there is precedence to provide supplemental antibiograms that highlight specific units of the hospital (e.g., ICU, emergency department, or outpatient clinics) or specific specimen sources (e.g., urine only). As noted earlier in this chapter, resistance rates in the ICU can differ from the general ward or emergency department [17, 42–44]. Understanding if a difference exists and communicating that to prescribers can be easily accomplished with an ICU-specific antibiogram. Studies have previously demonstrated that presenting data by ICU versus non-ICU could lead to different decisions on appropriate empirical antibiotic treatment [18, 45]. When separated by unit, certain uncommon species become infrequent and limit the ability to report an accurate susceptibility rate. CLSI recommends excluding groups that have fewer than 30 isolates per time period as the confidence interval around the estimate of susceptibility becomes larger as sample size decreases, especially in scenarios where low susceptibilities are observed. For example, with 30 *P. aeruginosa* collected from an ICU, a reported meropenem susceptibility of 70% has a 95% confidence interval of being between 50–85%. Alternatively, a footnote is recommended for reporting results of isolates with cumulative numbers lower than 30. The ICU antibiogram may also be a place to report the presence or rates of specific clinically significant phenotypes such as methicillin-resistant *Staphylococcus aureus* (MRSA), extended spectrum beta-lactamase (ESBL) producing *Enterobacteriaceae*, carbapenem-resistant Enterobacteriaceae (CRE), or the more recently described Difficult-to-Treat Resistance (DTR) measurement developed by the National Institutes of Health [46, 47].

Lastly, although it is not necessary to report detailed minimum inhibitory concentration (MIC) data on any antibiogram, the authors recommend that individual organism susceptibility reports from the microbiology lab should include not only categorical assessments (Susceptible, Susceptible-Dose-Dependent, Intermediate, Resistant), but also quantitative measures of potency (i.e., the MIC) for reported antibiotics in the ICU, if these data are available. MIC level data is welcomed and becomes essential for optimal dosing regimen selection (see Sect. "Antimicrobial Therapy" below), particularly in critically ill patients with severe infections [36, 48].

Antimicrobial Therapy

Empiric Therapy

The popular concept of "Getting it Right Up Front" with respect to antibiotic use stems from numerous studies of critically ill patients with severe infections [49]. From these studies, appropriate antibiotic therapy is defined as receipt of an antibiotic that has microbiological activity (i.e., the bacteria are susceptible to it) within 24–48 h of infection presentation. Timing is also critical, particularly in the most severe infection presentations such as septic shock. Kumar and colleagues demonstrated that each hour of delay in antimicrobial administration over the 6 h after presentation with hypotension was associated with an average decrease in survival of 7.6% [50]. These findings have been replicated in animal infection models [51]; furthermore, timing of appropriate antibiotic therapy has also been linked to duration of hospital stay [52]. As a result, the Surviving Sepsis Campaign guidelines recommend starting antimicrobials intravenously within 1 h of sepsis diagnosis [53].

Given the empirical nature of treating infections in the ICU setting, these findings "could" be extrapolated to prescribing the most broad-spectrum antibiotic or combinations of antibiotics to cover the most pathogens possible as quickly as possible. This concept is not unlike the view of Dr. Paul Ehrlich from over 100 years ago, where he suggested chemotherapy should be used to "Hit Hard and Hit Fast" [54]. Unfortunately, the above findings are largely in contrast with contemporary antibiotic stewardship tactics, where the most broad-spectrum, potent, and/or sometimes toxic antibiotics are reserved until clinically required. This balance of providing timely appropriate antibiotic quickly to the sickest patients while avoiding use of antibiotics unlikely to provide increased efficacy remains a significant challenge to ASP in the ICU setting.

Understanding the pathogens likely to cause the most common infections in an individual ICU and local resistance patterns should be key to achieving the above balance [36]. In many ICUs, combination therapy (e.g., vancomycin plus beta-lactam plus second antibiotic with activity against *P. aeruginosa)* can be utilized to increase the likelihood of having at least one antibiotic therapy prescribed empirically that will have in vitro activity. For instance, current IDSA guidelines for the treatment of ventilator-associated pneumonia recommend empirically adding an antibiotic with activity against MRSA as well as two active drugs for *P. aeruginosa* if rates of resistance are greater than 10–20% in the ICU [9]. While the 10–20% is admittedly arbitrary, most ICUs will meet these thresholds, justifying the need for empiric combination therapy. Similar recommendations are made by the Surviving Sepsis Campaign [53]. An alternative approach to combination therapy, at least against *P. aeruginosa*, would be to use a single broad-spectrum, potent agent that demonstrates equal or improved susceptibility when compared with traditional combination regimens. For example, when tested against 359 *P. aeruginosa* collected from ICUs in the USA, ceftolozane/tazobactam susceptibility was 95% compared with 88% and 71.6% by the next highest antibiotics, tobramycin and cefepime, respectively [55]. While combination therapy with tobramycin plus any of the anti-pseudomonas beta-lactams (cefepime, ceftazidime, meropenem, piperacillin/tazobactam) increased the likelihood of at least one drug being active to greater than 90%, no combination was still as high as ceftolozane/tazobactam monotherapy. For ICUs that harbor *P. aeruginosa* with significant rates of aminoglycoside or fluoroquinolone resistance, such a single drug approach may be a suitable option for consideration. Once understood, critical pathways or protocols may be used to guide ICU prescribers with initial selection of empiric antibiotic therapy as described earlier in this chapter.

After successful empiric therapy has been selected and identification and susceptibility of the pathogen reported, antibacterial therapy can be narrowed to a single antibiotic with activity against the organism. Garnacho-Montero and colleagues evaluated monotherapy compared with combination antibiotic therapy for the management of *P. aeruginosa* VAP in five ICUs [56]. While monotherapy was associated with an increase in inappropriate antibiotic therapy (90.5% vs. 56.7%, $p < 0.001$) and mortality (72.5% vs. 23.1%, $p < 0.05$), there was no difference in clinical endpoints for patients who received active antibiotic therapy with one versus two agents, suggesting the role for early combination therapy to maximize likelihood of appropriate antibiotic therapy followed by rapid de-escalation to a single active antibiotic. Based on this study and others, the VAP guidelines recommend de-escalation to active monotherapy once identification and susceptibility data are available [9]. Early combination therapy may also benefit patients with septic shock. In a propensity score matched cohort of 4662 patients with culture-positive septic shock, combination therapy with two active antibiotics against the Gram-positive or Gram-negative pathogen was associated with significant reductions in ICU (35.7% vs. 28.8%; odds ratio, 0.75; 95% confidence interval, 0.63–0.92; $p = 0.0006$) and hospital mortality (47.8% vs. 37.4%; odds ratio, 0.69; 95% confidence interval, 0.59–0.81; $p < 0.0001$) [57]. In addition to early therapy with two or more antibiotics, there may be individual ICU patients for whom combination therapy is justified for the full course. These challenging cases include multidrug resistant (MDR) organisms such as carbapenemase producing Enterobacteriaceae (CPE), *Acinetobacter baumannii*, and bacteremic pneumonia caused by MDR *P. aeruginosa* [58–63]. For these pathogens, combination therapy may even include one drug that is not active against the MDR organism isolated (e.g., meropenem plus colistin against CPE) because, while susceptibility may not appear in vitro, the inoculum size or pathogen subspecies may still allow for sufficient in vivo efficacy. Once identified, the ASP team, or more likely, the Infectious Diseases consult service, can assist the critical team with optimizing treatment regimens for these pathogens on an individual basis.

Altered Pharmacokinetics

It is well established that critically ill patients in the ICU may have significantly altered antibiotic pharmacokinetics that could directly affect achievable exposure [14]. A hyperdynamic state during sepsis could increase cardiac output resulting in augmented renal clearance (ARC) and reduced plasma concentrations. Altered fluid balances due to third spacing, altered protein binding, or both can lead to increases in volume of distribution. In contrast, renal and hepatic dysfunction may cause reductions in clearance thereby increasing plasma concentrations and requiring appropriate dosage reduction. Finally, numerous organ support modalities including continuous renal replacement therapy (CRRT) and extracorporeal membrane oxygenation (ECMO) can alter volumes of distribution and clearance resulting in variable concentrations. These challenges, combined with the potential for MDR pathogens with higher MICs in the ICU setting, make it even more important for the ASP team to confirm an optimal dose has been prescribed in the ICU.

Pharmacodynamic Optimization

The optimal dose of an antibiotic is one that attains its requisite exposure at the site of infection relative to its MIC against the causative pathogen(s). Table 8.2 provides a summary of various pharmacodynamic targets for currently available antibiotic classes. We refer the reader to the classic review on antibacterial pharmacodynamics by William Craig, MD [64].

Table 8.2 Summary of pharmacodynamics parameters and categorization of antibiotics. Adapted from Connors et al. [65]

Parameter	Description	Antibiotics and target exposure
Time above MIC (T > MIC)	Amount of time the antibiotic concentration is higher than the MIC, which can be expressed as the percent of the dosing interval that free drug concentrations remain above the MIC (*f*T > MIC) Often administered frequently, though an alternative strategy is to administer over a prolonged or continuous infusion	Penicillins: 50% *f*T > MIC Cephalosporins: 50–70% *f*T > MIC Monobactams: 50% *f*T > MIC Carbapenems: 30–40% *f*T > MIC
Maximum concentration to MIC ratio (C_{max}:MIC)	Ratio between the maximum concentration during each dosing interval and the MIC	Aminoglycosides: C_{max}:MIC ≥ 10–12
Area under the curve to MIC ratio (AUC:MIC)	The cumulative antibiotic exposure a patient receives relative to the MIC Can be expressed as total AUC:MIC or *f*AUC:MIC depending on specific studies	Fluoroquinolones Gram-negatives, total AUC:MIC > 125 Gram-positives, *f*AUC:MIC > 30–50 Vancomycin: Total AUC:MIC ≥ 400 Linezolid: Total AUC:MIC > 110 Polymyxins: *f*AUC/MIC > 12–15

In brief, concentration-dependent killing antibiotics exhibit antibacterial effects based on their maximum concentration relative to the MIC of the pathogen (Fig. 8.1). As the concentration increases relative to the MIC, the rate and extent of kill increases. Applicable pharmacodynamic indices for concentration-dependent killing antibiotics are the ratio between the free (i.e., unbound) maximum concentration and the MIC (C_{max}:MIC), the free area under the curve relative to the MIC (AUC:MIC), or both. Aminoglycosides are the iconic class of concentration-dependent antibiotics and plasma concentration monitoring is a familiar task associated with administration of its agents. The pharmacodynamic target for these agents is a C_{max}:MIC $\geq$ 10–12 and they can be administered as a high-dose, extended interval regimen to maximize killing while minimizing nephrotoxicity [65, 66]. Notably, the availability of gentamicin, tobramycin, and amikacin assays in many hospitals permits therapeutic drug monitoring (TDM) and individualization of dosing regimens for ICU patients based on observed concentrations and the MIC of the pathogen. Fluoroquinolones also display concentration-dependent killing; however, due to increases in adverse events associated with high peak concentrations, larger doses cannot routinely be adopted to address high MICs. Considering the target necessary to be bactericidal (AUC:MIC > 125), fluoroquinolone regimens must maximize exposure by balancing dosage and dosing frequency [65]. Should the MIC of the infective pathogen be too high to safely meet the target AUC:MIC, alternative or combination therapy is recommended.

In contrast, time-dependent killing antibiotics rely on concentrations remaining above the MIC for a proportion of each dosing interval to maximize antibacterial effects. This is best described by the percent of the dosing interval that unbound concentrations remain above the MIC (*f*T > MIC) (Fig. 8.1). Beta-lactam antibiotics including penicillins, cephalosporins, monobactams, and carbapenems are well-known for their time-dependent killing profiles. Their *f*T > MIC targets are ~50%, 50–70%, ~50%, and 30–40%, respectively [65]. Notably, although some

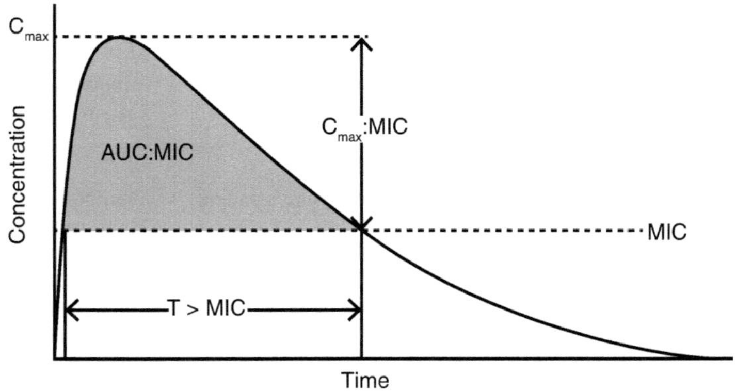

Fig. 8.1 Pharmacodynamic indices associated with antimicrobial activity. C_{max}/MIC, maximum concentration to minimum inhibitory concentration (MIC); AUC:MIC, area under the curve to MIC; T > MIC, time above the MIC

pharmacodynamic studies in the ICU setting have applied *f*T > MIC targets of 100% or even 100% T > 4xMIC [67], limited data are available to justify the need for these higher beta-lactam exposures [48]. Because it is important to maintain concentrations above the MIC for beta-lactams, continuous (i.e., 24 h) and prolonged (i.e., 3–4 h) infusions have become popular methods of administration that optimize *f*T > MIC, particularly against higher MICs [14, 48]. A combination of higher doses with continuous or prolonged infusions enables optimal exposure attainment against MICs that are even above the susceptibility breakpoint (Table 8.3) [68]. Therefore, to ideally optimize antibiotic pharmacodynamics in ICU patients, the laboratory defined MIC is a critical piece of information. Stewardship teams should work their microbiology labs to determine feasibility in making MICs available for certain organisms collected in the ICU.

It is still debated as to whether prolonged or continuous infusion regimens are superior for the beta-lactams [69]. The BLING (beta-lactam infusion group) I and BLING II randomized controlled trials compared continuous infusion with traditional infusion of the same beta-lactams for patients with severe sepsis. While BLING I [70], the smaller of the trials with only 60 patients, identified a significantly higher clinical cure rate (70% vs. 43%, $p = 0.037$) for patients receiving continuous infusion, the larger BLING II study was unable to recapitulate these observations [71]. Notably, few pathogens were identified in BLING II, and most were highly susceptible with low MICs, implying that the traditional infusion regimens also likely obtained optimal *f*T > MIC exposures. Despite the BLING II study findings, numerous meta-analyses have concluded that prolonged or continuous

Table 8.3 Maximum treatable MICs of traditional 0.5 h, prolonged (3 h), and continuous infusion beta-lactam regimens based on Monte Carlo simulation in patients with normal renal function (creatinine clearance 50–120 mL/min). Adapted from Monogue et al. [68]

Antibiotic	Regimen (infusion time)	Maximum treatable MIC (mg/L)
Ceftazidime	2 g Q8H (0.5 h)	4–8
	2 g Q8H (3 h)	8–16
	6 g Q24H (continuous)	8–16
Cefepime	2 g Q8H (0.5 h)	4–8
	2 g Q8H (3 h)	8–16
	6 g Q24H (continuous)	8–16
Imipenem	500 mg Q6H (0.5 h)	2
	500 mg Q6H (3 h)	4
Meropenem	500 mg Q6H (0.5 h)	2
	500 mg Q6H (3 h)	4
	1 g Q8H (0.5 h)	2
	1 g Q8H (3 h)	4–8
	2 g Q8H (0.5 h)	4
	2 g Q8H (3 h)	8–16
Piperacillin/tazobactam	4.5 g Q6H (0.5 h)	8
	4.5 g Q6H (3 h)	16–32
	18 g Q24H (continuous)	16–32

infusions for beta-lactams result in mortality reductions, specifically for patients in the ICU [72, 73]. In theory, the changes in pharmacokinetics among ICU patients combined with risks for higher MIC organisms provide an advantage to the pharmacodynamically optimized regimens.

Finally, vancomycin is a time-dependent killing antibiotic in which AUC/MIC has been linked to efficacy in patients with bloodstream infections caused by MRSA [74, 75]. Although routine trough monitoring has been recommended for decades, more recent clinical studies suggest that AUC/MIC is better linked to survival and AUC itself is predictive of nephrotoxicity. Vancomycin dosing regimens designed to achieve an AUC/MIC $\geq$ 400 instead of only targeting troughs >15 mg/L result in lower grams of vancomycin daily, fewer blood samples for vancomycin TDM, reduced rates of nephrotoxicity, and shorter length of stay, while providing similar clinical efficacy [76]. The 2020 IDSA and ASHP consensus guidelines on vancomycin recommend AUC-based monitoring for vancomycin [77]. To do this properly, at least two concentration values within the dosing interval are required to characterize AUC, or the ICU pharmacist/stewardship team could implement Bayesian software [14] to aid in estimating AUC off of a single plasma concentration. The utility of AUC:MIC monitoring for empiric vancomycin or for non-Staphylococcus aureus infections has yet to be determined. Due to increased safety and lower dosing this strategy may be preferred for patients on prolonged regimens of vancomycin; however, it may not be necessary for patients receiving short course therapy or are likely to be de-escalated from vancomycin prior to levels being collected. Importantly for ICU patients, loading doses are also required to achieve steady-state exposures quickly in this population.

Diagnostics

Diagnostic stewardship is a relatively new term that describes the role that laboratory plays in ASP programs. Indeed, there are more diagnostics approved annually targeting Infectious Diseases than there are new antibiotics. A couple of these diagnostic tools are worth highlighting for their role in the ICU setting.

Cultures are a mainstay of the diagnostic toolkit, especially when there is a clear source of infection. However, speed has become an issue due to the introduction of tools such as real-time polymerase chain reaction (PCR) and Matrix Assisted Laser Desorption Ionization Time-Of-Flight (MALDI-TOF) [78]. MALDI-TOF has already shown to speed time to pathogen recognition and decrease both mortality and length of stay in the ICU [79]. Although not performed in an ICU setting, Banerjee and colleagues compared outcomes associated with standard blood culture processing (control group), rapid multiplex PCR (rmPCR), and ASP-audited rmPCR. Compared with the control group the ASP-audited rapid multiplex PCR group had decreased broad-spectrum piperacillin-tazobactam use (p = 0.01), increased narrow-spectrum beta-lactam use (p = 0.04), and shortened the time to the appropriate intervention (de-escalation p < 0.001; escalation p = 0.04) without affecting mortality, length of stay, or cost [80]. These alternative methods would likely be most useful in ICUs with high rates of resistance.

Another challenge faced specifically by blood cultures is the potential for artificial sterilization that even antibiotic-binding resins may not sufficiently remedy. Carbapenems and cefepime, in particular, appear adept at sterilizing susceptible Gram-negatives in bottle culture bottles, regardless of the presence of antibiotic-binding resins [81]. Upwards of 84% of ICU patients received an antibiotic within 4 h prior to culture collection in one study [82]. Similar rates of prior antibiotic receipt were observed in our ICU (unpublished data). Therefore, in hospital ICUs that frequently use carbapenems or cefepime empirically, ICU patients with negative blood cultures at 24 h deserve a review of antibiotic timing relative to that of blood culture collection. In such cases that one of these antibiotics was received within 4 h prior to blood culture collection, repeat culture or the use of alternative diagnostics unlikely to be affected by prior antibiotic, such as T2 magnetic resonance technology, is recommended [83].

Biomarkers also serve to inform decisions on when to discontinue therapy. Procalcitonin (PCT), especially, has gained traction due to its ability to differentiate between infectious (including bacterial versus viral) and noninfectious syndromes, as well as the severity of the infection [84]. PCT, an amino peptide produced by the thyroid, is generally not detectable in healthy patients because it is typically converted to calcitonin before secretion into the bloodstream for participation in calcium homeostasis [85, 86]. However, extrathyroidal calcitonin can be generated and released during systemic inflammation [87, 88]. Of note, this occurs without an accompanying rise in calcitonin. Infectious inflammatory syndromes, in particular, produce significantly more PCT than noninfectious ones, theoretically making it an attractive parameter to monitor for both diagnostic and treatment response purposes [89–91].

Using PCT in the ICU to identify infectious cases from other inflammatory syndromes has its challenges. A number of studies have observed that a PCT threshold of 1–1.1 ng/mL bore sensitivity rates of 77–97% and specificity rates of 78–89% [91–93]. A number of factors contribute to these wide ranges. It can take anywhere from 24 to 48 h for PCT to rise due to infection, so initial measurements may not reflect the "true" value that would be useful for diagnosis. PCT's continued presence thereafter may also convolute attempts to diagnose any subsequent infections. Additionally, localized infections do not trigger the PCT increase seen from systemic infections, further narrowing its utility as a diagnostic biomarker [94]. It is not surprising, then, that studies focusing on PCT as an indicator to initiate antibiotic therapy have remained largely discouraging. The PRORATA trial, a large randomized controlled ICU trial, tested an algorithm that discouraged initiation when PCT was <0.5 ng/mL. It observed similar initiation rates between the PCT algorithm and control groups, though more than one-fifth of those in the PCT group still received antibiotics even when the algorithm discouraged it [95]. Layios and colleagues reported similar observations using the same PCT cutoff with over half of patients with concentrations below it treated against the algorithm's recommendation [96].

Measuring and trending PCT in order to halt antibiotics, on the other hand, is much more reliable. Schroeder and colleagues investigated the use of a PCT-guided algorithm in deciding when to discontinue antibiotics in surgical ICU patients with severe sepsis post-abdominal surgery. In the PCT group, once the patient had improved clinically and PCT had fallen to 1 ng/mL or less or decreased by at least 25–35% of the original PCT reading over three consecutive days, antibiotic therapy was discontinued. The algorithm resulted in a statistically significant decrease in duration of antibiotic treatment (6.6 ± 1.1 days vs. 8.3 ± 0.7 days, $p < 0.001$), as well as a 17.8% decrease in antibiotic costs [84]. A similar study was run in Switzerland where antibiotics were discontinued if PCT fell by at least 90% by the third or fifth day of treatment, depending on what the initial PCT value was. This study also saw decreases in antibiotic therapy, as well as a shorter stay in the ICU [97]. The PRORATA trial mentioned earlier also looked at antibiotic discontinuation and found that the PCT group had significantly shortened durations of therapy (14.3 ± 9.1 vs. 11.6 ± 8.2 days, $p < 0.0001$). Subgroup analysis yielded similar results across disease severity and site of infection. Other clinical endpoints—mortality on Days 28 and 60, infection relapse, superinfection, and development of a multidrug-resistant infection—were also similar between the PCT and control groups [95]. Many of the durations reported in PCT trials extend beyond what is now routine recommended for many ICU infections. With recent randomized clinical trials and clinical practice guidelines advocating for shorter course therapy the cost and use of PCT must continue to be evaluated in the real-world setting.

Perhaps the most challenging use of diagnostics for ICU patients and stewardship is trusting in their negative results to discontinue antibiotics. Due to the use of broad-spectrum antibiotics empirically and the severity of illness, clinicians may be reluctant to discontinue or narrow antibiotic therapy when a patient is improving on broad-spectrum combination therapy. As described above, PCT may play a valuable role in demonstrating response in order to justify concluding antibiotic therapy. But what about early antibiotic therapy when diagnostics suggest no bacterial infection is present? Raman and colleagues studied the effects of early antibiotic discontinuation in ICU patients with clinically suspected ventilator-associated pneumonia and negative quantitative bronchoscopy cultures (<10^4 CFU/mL) [98]. Discontinuation was deemed early if it was done within one calendar day of final negative culture report and late if it was done at any other time after that. Mortality did not differ between the two groups ($p = 0.642$), and duration of therapy did not differ between patients that survived and those that did not ($p = 0.87$). The early discontinuation also seemed to confer the benefit of fewer superinfections (22.5% vs. 42.9%, $p = 0.008$). Even the phrasing of negative culture reports can have a substantial impact [99], which is an avenue that should be explored if current practices may be contributing to excessively prolonged treatment.

Negative nasal MRSA PCR screening results in pneumonia patients are another useful diagnostic for de-escalation of empirical MRSA therapy because they strongly suggest the absence of MRSA in the lower respiratory tract [100]. Studies investigating pharmacist-driven MRSA PCR nasal swab protocols have demonstrated their value through retrospective analyses between pre- and post-protocol groups. Baby and colleagues observed that the protocol shortened the mean duration of vancomycin or linezolid therapy by 46.6 h ($p < 0.0001$) and decreased the percentage of patients requiring vancomycin levels or dose modifications ($p = 0.02$)

without changing days to clinical improvement, length of stay, or hospital mortality [101]. Dadzie and colleagues reported shortened vancomycin therapy ($p = 0.044$) and shortened length of stay ($p = 0.006$) without affecting clinical outcomes [102].

Lastly, it is vital to consider whether specific diagnostic tests are required at all. Tests should only be ordered when their results may affect treatment. For instance, viral respiratory panels have gained traction for use in the ICU on pneumonia patients. Theoretically, this could be a strategy to reduce antibiotic use in cases where the causative pathogen is indeed a virus. However, recent studies have observed that implementation of these viral panels has been unsuccessful in reducing inappropriate antibiotic management of viral infection [103, 104]. While more information may seem preferable, unnecessary testing brings a host of issues in addition to wasted healthcare system resources. At best, the results are an innocuous addition to the patient's chart, but they can also be used to support unnecessary treatment, such as in cases of asymptomatic bacteriuria [105, 106].

Summary

By nature, the critically ill patients found in the ICU may benefit a great deal from the services offered by an ASP. The severity of the situation demands timely and accurate diagnostics and evidence-based empiric therapy until more information is obtained. Their altered pharmacokinetic profiles require individualized dosing, potentially with therapeutic drug monitoring, and alternatively, pharmacodynamically-driven regimens that take advantage of the drug's intrinsic properties. The ICU setting itself already attracts patients with more resistant pathogens, making it all the more important to implement formulary restrictions, to encourage de-escalation and discontinuation of antimicrobials as soon as is clinically advisable, and to develop and maintain an ICU-specific antibiogram. This collaboration between the ICU team, ID specialists, and the microbiology department is the clear option for both patients and the institutions that care for them.

References

1. Halpern NA, Pastores SM. Critical care medicine in the United States 2000-2005: an analysis of bed numbers, occupancy rates, payer mix, and costs. Crit Care Med. 2010;38(1):65–71.
2. Mullins PM, Goyal M, Pines JM. National growth in intensive care unit admissions from emergency departments in the United States from 2002 to 2009. Acad Emerg Med. 2013;20(5):479–86.
3. Sjoding MW, Prescott HC, Wunsch H, Iwashyna TJ, Cooke CR. Longitudinal changes in ICU admissions among elderly patients in the United States. Crit Care Med. 2016;44(7):1353–60.
4. Magill SS, Edwards JR, Bamberg W, Beldavs ZG, Dumyati G, Kainer MA, et al. Multistate point-prevalence survey of health care-associated infections. N Engl J Med. 2014;370(13):1198–208.
5. Donowitz LG, Wenzel RP, Hoyt JW. High risk of hospital-acquired infection in the ICU patient. Crit Care Med. 1982;10(6):355–7.

6. Fagon JY, Chastre J, Hance AJ, Montravers P, Novara A, Gibert C. Nosocomial pneumonia in ventilated patients: a cohort study evaluating attributable mortality and hospital stay. Am J Med. 1993;94(3):281–8.
7. Rello J, Ollendorf DA, Oster G, Vera-Llonch M, Bellm L, Redman R, et al. Epidemiology and outcomes of ventilator-associated pneumonia in a large US database. Chest. 2002;122(6):2115–21.
8. Chastre J, Fagon JY. Ventilator-associated pneumonia. Am J Respir Crit Care Med. 2002;165(7):867–903.
9. Kalil AC, Metersky ML, Klompas M, Muscedere J, Sweeney DA, Palmer LB, et al. Management of adults with hospital-acquired and ventilator-associated pneumonia: 2016 clinical practice guidelines by the Infectious Diseases Society of America and the American Thoracic Society. Clin Infect Dis. 2016;63(5):e61–e111.
10. Martin GS. Sepsis, severe sepsis and septic shock: changes in incidence, pathogens and outcomes. Expert Rev Anti-Infect Ther. 2012;10(6):701–6.
11. Vincent JL, Rello J, Marshall J, Silva E, Anzueto A, Martin CD, et al. International study of the prevalence and outcomes of infection in intensive care units. JAMA. 2009;302(21):2323–9.
12. Luyt CE, Brechot N, Trouillet JL, Chastre J. Antibiotic stewardship in the intensive care unit. Crit Care. 2014;18(5):480.
13. Micek ST, Heard KM, Gowan M, Kollef MH. Identifying critically ill patients at risk for inappropriate antibiotic therapy: a pilot study of a point-of-care decision support alert. Crit Care Med. 2014;42(8):1832–8.
14. Roberts JA, Abdul-Aziz MH, Lipman J, Mouton JW, Vinks AA, Felton TW, et al. Individualised antibiotic dosing for patients who are critically ill: challenges and potential solutions. Lancet Infect Dis. 2014;14(6):498–509.
15. Douglas IS. New diagnostic methods for pneumonia in the ICU. Curr Opin Infect Dis. 2016;29(2):197–204.
16. Chiotos K, Tamma PD, Gerber JS. Antibiotic stewardship in the intensive care unit: challenges and opportunities. Infect Control Hosp Epidemiol. 2019;40(6):693–8.
17. Eagye KJ, Banevicius MA, Nicolau DP. Pseudomonas aeruginosa is not just in the intensive care unit any more: implications for empirical therapy. Crit Care Med. 2012;40(4):1329–32.
18. Binkley S, Fishman NO, LaRosa LA, Marr AM, Nachamkin I, Wordell D, et al. Comparison of unit-specific and hospital-wide antibiograms: potential implications for selection of empirical antimicrobial therapy. Infect Control Hosp Epidemiol. 2006;27(7):682–7.
19. Barlam TF, Cosgrove SE, Abbo LM, MacDougall C, Schuetz AN, Septimus EJ, et al. Implementing an antibiotic stewardship program: guidelines by the Infectious Diseases Society of America and the Society for Healthcare Epidemiology of America. Clin Infect Dis. 2016;62(10):e51–77.
20. Dellit TH, Owens RC, McGowan JE Jr, Gerding DN, Weinstein RA, Burke JP, et al. Infectious Diseases Society of America and the Society for Healthcare Epidemiology of America guidelines for developing an institutional program to enhance antimicrobial stewardship. Clin Infect Dis. 2007;44(2):159–77.
21. Abdallah M, Badawi M, Amirah MF, Rasheed A, Mady AF, Alodat M, et al. Impact of carbapenem restriction on the antimicrobial susceptibility pattern of Pseudomonas aeruginosa isolates in the ICU. J Antimicrob Chemother. 2017;72(11):3187–90.
22. Ntagiopoulos PG, Paramythiotou E, Antoniadou A, Giamarellou H, Karabinis A. Impact of an antibiotic restriction policy on the antibiotic resistance patterns of gram-negative microorganisms in an intensive care unit in Greece. Int J Antimicrob Agents. 2007;30(4):360–5.
23. Aubert G, Carricajo A, Vautrin AC, Guyomarc'h S, Fonsale N, Page D, et al. Impact of restricting fluoroquinolone prescription on bacterial resistance in an intensive care unit. J Hosp Infect. 2005;59(2):83–9.
24. Brahmi N, Blel Y, Kouraichi N, Lahdhiri S, Thabet H, Hedhili A, et al. Impact of ceftazidime restriction on gram-negative bacterial resistance in an intensive care unit. J Infect Chemother. 2006;12(4):190–4.

25. Meyer E, Lapatschek M, Bechtold A, Schwarzkopf G, Gastmeier P, Schwab F. Impact of restriction of third generation cephalosporins on the burden of third generation cephalosporin resistant K. pneumoniae and E. coli in an ICU. Intensive Care Med. 2009;35(5):862–70.
26. Rahal JJ, Urban C, Horn D, Freeman K, Segal-Maurer S, Maurer J, et al. Class restriction of cephalosporin use to control total cephalosporin resistance in nosocomial Klebsiella. JAMA. 1998;280(14):1233–7.
27. O'Leary EN, van Santen KL, Webb AK, Pollock DA, Edwards JR, Srinivasan A. Uptake of antibiotic stewardship programs in US acute care hospitals: findings from the 2015 National Healthcare Safety Network Annual Hospital Survey. Clin Infect Dis. 2017;65(10):1748–50.
28. Nhan D, Lentz EJM, Steinberg M, Bell CM, Morris AM. Structure of antimicrobial stewardship programs in leading US hospitals: findings of a Nationwide survey. Open Forum Infect Dis. 2019;6(4):ofz104.
29. Tamma PD, Avdic E, Keenan JF, Zhao Y, Anand G, Cooper J, et al. What is the more effective antibiotic stewardship intervention: preprescription authorization or postprescription review with feedback? Clin Infect Dis. 2017;64(5):537–43.
30. Lindsay PJ, Rohailla S, Taggart LR, Lightfoot D, Havey T, Daneman N, et al. Antimicrobial stewardship and intensive care unit mortality: a systematic review. Clin Infect Dis. 2019;68(5):748–56.
31. Morris AM, Bai A, Burry L, Dresser LD, Ferguson ND, Lapinsky SE, et al. Long-term effects of phased implementation of antimicrobial stewardship in academic ICUs: 2007-2015. Crit Care Med. 2019;47(2):159–66.
32. Abbara S, Pitsch A, Jochmans S, Hodjat K, Cherrier P, Monchi M, et al. Impact of a multimodal strategy combining a new standard of care and restriction of carbapenems, fluoroquinolones and cephalosporins on antibiotic consumption and resistance of Pseudomonas aeruginosa in a French intensive care unit. Int J Antimicrob Agents. 2019;53(4):416–22.
33. Nicasio AM, Eagye KJ, Nicolau DP, Shore E, Palter M, Pepe J, et al. Pharmacodynamic-based clinical pathway for empiric antibiotic choice in patients with ventilator-associated pneumonia. J Crit Care. 2010;25(1):69–77.
34. Wilde AM, Nailor MD, Nicolau DP, Kuti JL. Inappropriate antibiotic use due to decreased compliance with a ventilator-associated pneumonia computerized clinical pathway: implications for continuing education and prospective feedback. Pharmacotherapy. 2012;32(8):755–63.
35. van Duijn PJ, Verbrugghe W, Jorens PG, Spohr F, Schedler D, Deja M, et al. The effects of antibiotic cycling and mixing on antibiotic resistance in intensive care units: a cluster-randomised crossover trial. Lancet Infect Dis. 2018;18(4):401–9.
36. Kuti JL, Shore E, Palter M, Nicolau DP. Tackling empirical antibiotic therapy for ventilator-associated pneumonia in your ICU: guidance for implementing the guidelines. Semin Respir Crit Care Med. 2009;30(1):102–15.
37. Ibrahim EH, Ward S, Sherman G, Schaiff R, Fraser VJ, Kollef MH. Experience with a clinical guideline for the treatment of ventilator-associated pneumonia. Crit Care Med. 2001;29(6):1109–15.
38. Arnold HM, Micek ST, Skrupky LP, Kollef MH. Antibiotic stewardship in the intensive care unit. Semin Respir Crit Care Med. 2011;32(2):215–27.
39. Kollef MH, Micek ST. Antimicrobial stewardship programs: mandatory for all ICUs. Crit Care. 2012;16(6):179.
40. Gilbert DN. Influence of an infectious diseases specialist on ICU multidisciplinary rounds. Crit Care Res Pract. 2014;2014:307817.
41. Trupka T, Fisher K, Micek ST, Juang P, Kollef MH. Enhanced antimicrobial de-escalation for pneumonia in mechanically ventilated patients: a cross-over study. Crit Care. 2017;21(1):180.
42. Prabaker K, Weinstein RA. Trends in antimicrobial resistance in intensive care units in the United States. Curr Opin Crit Care. 2011;17(5):472–9.
43. Puzniak L, DePestel DD, Srinivasan A, Ye G, Murray J, Merchant S, et al. A combination antibiogram evaluation for Pseudomonas aeruginosa in respiratory and blood sources from Intensive Care Unit (ICU) and non-ICU settings in U.S. Hospitals. Antimicrob Agents Chemother. 2019;63(4):e02564-18.

44. Asempa TE, Nicolau DP, Kuti JL. Carbapenem non-susceptible Pseudomonas aeruginosa from intensive care units in the United States: a potential role for new beta-lactam combination agents. J Clin Microbiol. 2019;57:e00535-19.
45. Green DL. Selection of an empiric antibiotic regimen for hospital-acquired pneumonia using a unit and culture-type specific antibiogram. J Intensive Care Med. 2005;20(5):296–301.
46. Kadri SS, Adjemian J, Lai YL, Spaulding AB, Ricotta E, Prevots DR, et al. Difficult-to-treat resistance in gram-negative bacteremia at 173 US hospitals: retrospective cohort analysis of prevalence, predictors, and outcome of resistance to all first-line agents. Clin Infect Dis. 2018;67(12):1803–14.
47. Kadri SS, Lai YLE, Ricotta EE, Strich JR, Babiker A, Rhee C, et al. External validation of difficult-to-treat resistance prevalence and mortality risk in gram-negative bloodstream infection using electronic health record data from 140 US hospitals. OPem Forum Infect Dis. 2019;6(4):ofz110.
48. Grupper M, Kuti JL, Nicolau DP. Continuous and prolonged intravenous beta-lactam dosing: implications for the clinical laboratory. Clin Microbiol Rev. 2016;29(4):759–72.
49. Kollef MH. Broad-spectrum antimicrobials and the treatment of serious bacterial infections: getting it right up front. Clin Infect Dis. 2008;47(Suppl 1):S3–13.
50. Kumar A, Roberts D, Wood KE, Light B, Parrillo JE, Sharma S, et al. Duration of hypotension before initiation of effective antimicrobial therapy is the critical determinant of survival in human septic shock. Crit Care Med. 2006;34(6):1589–96.
51. Kumar A, Haery C, Paladugu B, Kumar A, Symeoneides S, Taiberg L, et al. The duration of hypotension before the initiation of antibiotic treatment is a critical determinant of survival in a murine model of Escherichia coli septic shock: association with serum lactate and inflammatory cytokine levels. J Infect Dis. 2006;193(2):251–8.
52. Zhang D, Micek ST, Kollef MH. Time to appropriate antibiotic therapy is an independent determinant of postinfection ICU and hospital lengths of stay in patients with sepsis. Crit Care Med. 2015;43(10):2133–40.
53. Rhodes A, Evans LE, Alhazzani W, Levy MM, Antonelli M, Ferrer R, et al. Surviving sepsis campaign: international guidelines for management of sepsis and septic shock: 2016. Crit Care Med. 2017;45(3):486–552.
54. Ehrlich P. Address in pathology, ON CHEMIOTHERAPY: delivered before the seventeenth international congress of medicine. Br Med J. 1913;2(2746):353–9.
55. Goodlet KJ, Nicolau DP, Nailor MD. In vitro comparison of Ceftolozane-Tazobactam to traditional Beta-lactams and Ceftolozane-Tazobactam as an alternative to combination antimicrobial therapy for Pseudomonas aeruginosa. Antimicrob Agents Chemother. 2017;61(12):e01350-17.
56. Garnacho-Montero J, Sa-Borges M, Sole-Violan J, Barcenilla F, Escoresca-Ortega A, Ochoa M, et al. Optimal management therapy for Pseudomonas aeruginosa ventilator-associated pneumonia: an observational, multicenter study comparing monotherapy with combination antibiotic therapy. Crit Care Med. 2007;35(8):1888–95.
57. Kumar A, Zarychanski R, Light B, Parrillo J, Maki D, Simon D, et al. Early combination antibiotic therapy yields improved survival compared with monotherapy in septic shock: a propensity-matched analysis. Crit Care Med. 2010;38(9):1773–85.
58. Bassetti M, Giacobbe DR, Giamarellou H, Viscoli C, Daikos GL, Dimopoulos G, et al. Management of KPC-producing Klebsiella pneumoniae infections. Clin Microbiol Infect. 2018;24(2):133–44.
59. Tumbarello M, Trecarichi EM, De Rosa FG, Giannella M, Giacobbe DR, Bassetti M, et al. Infections caused by KPC-producing Klebsiella pneumoniae: differences in therapy and mortality in a multicentre study. J Antimicrob Chemother. 2015;70(7):2133–43.
60. Durante-Mangoni E, Signoriello G, Andini R, Mattei A, De Cristoforo M, Murino P, et al. Colistin and rifampicin compared with colistin alone for the treatment of serious infections due to extensively drug-resistant Acinetobacter baumannii: a multicenter, randomized clinical trial. Clin Infect Dis. 2013;57(3):349–58.

61. Lopez-Cortes LE, Cisneros JM, Fernandez-Cuenca F, Bou G, Tomas M, Garnacho-Montero J, et al. Monotherapy versus combination therapy for sepsis due to multidrug-resistant Acinetobacter baumannii: analysis of a multicentre prospective cohort. J Antimicrob Chemother. 2014;69(11):3119–26.
62. Park SY, Park HJ, Moon SM, Park KH, Chong YP, Kim MN, et al. Impact of adequate empirical combination therapy on mortality from bacteremic Pseudomonas aeruginosa pneumonia. BMC Infect Dis. 2012;12:308.
63. Rigatto MH, Vieira FJ, Antochevis LC, Behle TF, Lopes NT, Zavascki AP. Polymyxin B in combination with antimicrobials lacking in vitro activity versus Polymyxin B in monotherapy in critically ill patients with Acinetobacter baumannii or Pseudomonas aeruginosa infections. Antimicrob Agents Chemother. 2015;59(10):6575–80.
64. Craig WA. Pharmacokinetic/pharmacodynamic parameters: rationale for antibacterial dosing of mice and men. Clin Infect Dis. 1998;26(1):1–10.
65. Connors KP, Kuti JL, Nicolau DP. Optimizing antibiotic pharmacodynamics for clinical practice. Pharmaceut Anal Acta. 2013;4(3):214.
66. Nicolau DP, Freeman CD, Belliveau PP, Nightingale CH, Ross JW, Quintiliani R. Experience with a once-daily aminoglycoside program administered to 2,184 adult patients. Antimicrob Agents Chemother. 1995;39(3):650–5.
67. Roberts JA, Paul SK, Akova M, Bassetti M, De Waele JJ, Dimopoulos G, et al. DALI: defining antibiotic levels in intensive care unit patients: are current beta-lactam antibiotic doses sufficient for critically ill patients? Clin Infect Dis. 2014;58(8):1072–83.
68. Monogue ML, Kuti JL, Nicolau DP. Optimizing antibiotic dosing strategies for the treatment of gram-negative infections in the era of resistance. Expert Rev Clin Pharmacol. 2016;9(3):459–76.
69. MacVane SH, Kuti JL, Nicolau DP. Prolonging beta-lactam infusion: a review of the rationale and evidence, and guidance for implementation. Int J Antimicrob Agents. 2014;43(2):105–13.
70. Dulhunty JM, Roberts JA, Davis JS, Webb SA, Bellomo R, Gomersall C, et al. Continuous infusion of beta-lactam antibiotics in severe sepsis: a multicenter double-blind, randomized controlled trial. Clin Infect Dis. 2013;56(2):236–44.
71. Dulhunty JM, Roberts JA, Davis JS, Webb SA, Bellomo R, Gomersall C, et al. A multicenter randomized trial of continuous versus intermittent beta-lactam infusion in severe sepsis. Am J Respir Crit Care Med. 2015;192(11):1298–305.
72. Teo J, Liew Y, Lee W, Kwa AL. Prolonged infusion versus intermittent boluses of beta-lactam antibiotics for treatment of acute infections: a meta-analysis. Int J Antimicrob Agents. 2014;43(5):403–11.
73. Rhodes NJ, Liu J, O'Donnell JN, Dulhunty JM, Abdul-Aziz MH, Berko PY, et al. Prolonged infusion piperacillin-Tazobactam decreases mortality and improves outcomes in severely ill patients: results of a systematic review and meta-analysis. Crit Care Med. 2018;46(2):236–43.
74. Zasowski EJ, Lodise TP. The importance of individualized vancomycin dosing to ensure optimal exposure early in therapy. J Clin Pharmacol. 2018;58(9):1131–3.
75. Lodise TP, Rosenkranz SL, Finnemeyer M, Evans S, Sims M, Zervos MJ, et al. The emperor's new clothes: prospective observational evaluation of the association between initial vancomycin exposure and failure rates among adult hospitalized patients with MRSA bloodstream infections (PROVIDE). Clin Infect Dis. 2020;70(8):1536–45.
76. Neely MN, Kato L, Youn G, Kraler L, Bayard D, van Guilder M, et al. Prospective trial on the use of trough concentration versus area under the curve to determine therapeutic vancomycin dosing. Antimicrob Agents Chemother. 2018;62(2):e02042-17.
77. Rybak MJ, Le J, Lodise TP, Levine DP, Bradley JS, Liu C, et al. Therapeutic monitoring of vancomycin for serious methicillin-resistant Staphylococcus aureus infections: a revised consensus guideline and review by the American Society of Health-System Pharmacists, the Infectious Diseases Society of America, the Pediatric Infectious Diseases Society, and the Society of Infectious Diseases Pharmacists. Am J Health Syst Pharm. 2020;77(11):835–64.
78. Johnson I, Banks V. Antibiotic stewardship in critical care. BJA Educ. 2017;17(4):111–6.

79. Huang AM, Newton D, Kunapuli A, Gandhi TN, Washer LL, Isip J, et al. Impact of rapid organism identification via matrix-assisted laser desorption/ionization time-of-flight combined with antimicrobial stewardship team intervention in adult patients with bacteremia and candidemia. Clin Infect Dis. 2013;57(9):1237–45.
80. Banerjee R, Teng CB, Cunningham SA, Ihde SM, Steckelberg JM, Moriarty JP, et al. Randomized trial of rapid multiplex polymerase chain reaction-based blood culture identification and susceptibility testing. Clin Infect Dis. 2015;61(7):1071–80.
81. Chen IH, Nicolau DP, Kuti JL. Recovery of gram-negative bacteria from aerobic blood culture bottles containing antibiotic binding resins after exposure to β-lactam and fluoroquinolone concentrations. J Clin Microbiol. 2019;57:e00849-19.
82. Zadroga R, Williams DN, Gottschall R, Hanson K, Nordberg V, Deike M, et al. Comparison of 2 blood culture media shows significant differences in bacterial recovery for patients on antimicrobial therapy. Clin Infect Dis. 2013;56(6):790–7.
83. Nguyen MH, Clancy CJ, Pasculle AW, Pappas PG, Alangaden G, Pankey GA, et al. Performance of the T2Bacteria panel for diagnosing bloodstream infections: a diagnostic accuracy study. Ann Intern Med. 2019;170:845.
84. Schroeder S, Hochreiter M, Koehler T, Schweiger AM, Bein B, Keck FS, et al. Procalcitonin (PCT)-guided algorithm reduces length of antibiotic treatment in surgical intensive care patients with severe sepsis: results of a prospective randomized study. Langenbecks Arch Surg. 2009;394(2):221–6.
85. Becker KL, Nylen ES, White JC, Muller B, Snider RH Jr. Clinical review 167: procalcitonin and the calcitonin gene family of peptides in inflammation, infection, and sepsis: a journey from calcitonin back to its precursors. J Clin Endocrinol Metab. 2004;89(4):1512–25.
86. Muller B, White JC, Nylen ES, Snider RH, Becker KL, Habener JF. Ubiquitous expression of the calcitonin-i gene in multiple tissues in response to sepsis. J Clin Endocrinol Metab. 2001;86(1):396–404.
87. Meisner M, Brunkhorst FM, Reith HB, Schmidt J, Lestin HG, Reinhart K. Clinical experiences with a new semi-quantitative solid phase immunoassay for rapid measurement of procalcitonin. Clin Chem Lab Med. 2000;38(10):989–95.
88. Wagner KE, Martinez JM, Vath SD, Snider RH, Nylen ES, Becker KL, et al. Early immunoneutralization of calcitonin precursors attenuates the adverse physiologic response to sepsis in pigs. Crit Care Med. 2002;30(10):2313–21.
89. Luyt CE, Combes A, Reynaud C, Hekimian G, Nieszkowska A, Tonnellier M, et al. Usefulness of procalcitonin for the diagnosis of ventilator-associated pneumonia. Intensive Care Med. 2008;34(8):1434–40.
90. Mofidi R, Suttie SA, Patil PV, Ogston S, Parks RW. The value of procalcitonin at predicting the severity of acute pancreatitis and development of infected pancreatic necrosis: systematic review. Surgery. 2009;146(1):72–81.
91. Muller B, Becker KL, Schachinger H, Rickenbacher PR, Huber PR, Zimmerli W, et al. Calcitonin precursors are reliable markers of sepsis in a medical intensive care unit. Crit Care Med. 2000;28(4):977–83.
92. Harbarth S, Holeckova K, Froidevaux C, Pittet D, Ricou B, Grau GE, et al. Diagnostic value of procalcitonin, interleukin-6, and interleukin-8 in critically ill patients admitted with suspected sepsis. Am J Respir Crit Care Med. 2001;164(3):396–402.
93. Wacker C, Prkno A, Brunkhorst FM, Schlattmann P. Procalcitonin as a diagnostic marker for sepsis: a systematic review and meta-analysis. Lancet Infect Dis. 2013;13(5):426–35.
94. Brechot N, Hekimian G, Chastre J, Luyt CE. Procalcitonin to guide antibiotic therapy in the ICU. Int J Antimicrob Agents. 2015;46(Suppl 1):S19–24.
95. Bouadma L, Luyt CE, Tubach F, Cracco C, Alvarez A, Schwebel C, et al. Use of procalcitonin to reduce patients' exposure to antibiotics in intensive care units (PRORATA trial): a multicentre randomised controlled trial. Lancet. 2010;375(9713):463–74.
96. Layios N, Lambermont B, Canivet JL, Morimont P, Preiser JC, Garweg C, et al. Procalcitonin usefulness for the initiation of antibiotic treatment in intensive care unit patients. Crit Care Med. 2012;40(8):2304–9.

97. Nobre V, Harbarth S, Graf JD, Rohner P, Pugin J. Use of procalcitonin to shorten antibiotic treatment duration in septic patients: a randomized trial. Am J Respir Crit Care Med. 2008;177(5):498–505.
98. Raman K, Nailor MD, Nicolau DP, Aslanzadeh J, Nadeau M, Kuti JL. Early antibiotic discontinuation in patients with clinically suspected ventilator-associated pneumonia and negative quantitative bronchoscopy cultures. Crit Care Med. 2013;41(7):1656–63.
99. Musgrove MA, Kenney RM, Kendall RE, Peters M, Tibbetts R, Samuel L, et al. Microbiology comment nudge improves pneumonia prescribing. Open Forum Infect Dis. 2018;5(7):ofy162.
100. Parente DM, Cunha CB, Mylonakis E, Timbrook TT. The clinical utility of methicillin-resistant Staphylococcus aureus (MRSA) nasal screening to rule out MRSA pneumonia: a diagnostic meta-analysis with antimicrobial stewardship implications. Clin Infect Dis. 2018;67(1):1–7.
101. Baby N, Faust AC, Smith T, Sheperd LA, Knoll L, Goodman EL. Nasal methicillin-resistant staphylococcus aureus (MRSA) PCR testing reduces the duration of MRSA-targeted therapy in patients with suspected MRSA Pneumonia. Antimicrob Agents Chemother. 2017;61(4):e02432-16.
102. Dadzie P, Dietrich T, Ashurst J. Impact of a pharmacist-driven methicillin-resistant Staphylococcus aureus polymerase chain reaction nasal swab protocol on the de-escalation of empiric vancomycin in patients with pneumonia in a rural healthcare setting. Cureus. 2019;11(12):e6378.
103. Kenes M, Maxam M, Thomas K. Respiratory viral infection testing in critically ill patients diagnosed with pneumonia. Crit Care Med. 2019;47(1):268.
104. Timbrook T, Maxam M, Bosso J. Antibiotic discontinuation rates associated with positive respiratory viral panel and low procalcitonin results in proven or suspected respiratory infections. Infect Dis Ther. 2015;4(3):297–306.
105. Leone M, Perrin AS, Granier I, Visintini P, Blasco V, Antonini F, et al. A randomized trial of catheter change and short course of antibiotics for asymptomatic bacteriuria in catheterized ICU patients. Intensive Care Med. 2007;33(4):726–9.
106. Lin G, Knowlson S, Nguyen H, Cooper K, Pryor RJ, Doll M, et al. Urine test stewardship for catheterized patients in the critical care setting: provider perceptions and impact of electronic order set interventions. Am J Infect Control. 2019;47:1277.

Antimicrobial Stewardship in Pediatric Patients

9

Jennifer E. Girotto and Nicholas Bennett

Introduction

The goals of antimicrobial stewardship programs (ASP) are the same in pediatric patients as they are in other populations: to ensure patients receive the narrowest effective antimicrobial therapy for the shortest effective duration. Further, ASPs have similar challenges in trying to curtail inappropriate antimicrobial usage. Current estimates suggest that 11–30% of pediatric outpatient and inpatient antibiotic prescriptions are inappropriate [1, 2]. This chapter will review the multiple factors and variables that should be incorporated into pediatric ASP including significant pharmacokinetic variability, distinct disease states and organisms, vaccine considerations, as well as limited data on newer agents and testing applicability.

Pharmacokinetic/Pharmacodynamic Considerations in Pediatrics

The pediatric population spans a significant age group from preterm neonates through adolescents, and one of the more impactful aspects of this variability is with regard to pharmacokinetics. Just after birth, babies have large total body water and extracellular fluid volumes, which result in relatively large volumes of distribution for many medications. This increased volume of distribution will decrease significantly throughout infancy and childhood. At the same time these newborns' renal clearances are very immature, especially in those born less than 35 weeks gestation.

J. E. Girotto (✉)
Connecticut Children's, University of Connecticut, Hartford, CT, USA
e-mail: jgirotto@connecticutchildrens.org

N. Bennett
ICON Plc, Dublin, Ireland

S. Doron, M. Campion (eds.), *Antimicrobial Stewardship in Non-Traditional Settings*, https://doi.org/10.1007/978-3-031-21710-4_9

The renal clearance increases after the neonatal period, with young children having extremely fast renal clearance of medications. Their clearance slowly decreases throughout childhood and adolescence. Even in adolescence, these patients will continue to have increased renal clearance versus healthy adult patients. This is important, as it makes it more challenging to obtain pharmacodynamic targets (e.g., time above the minimum inhibitory concentration (T > MIC), area under the curve (AUC) to MIC) with typical adult dosing.

For example, Monte Carlo simulations of antibiotic dosing in pediatric patients have shown that achieving bactericidal activity in children with beta-lactams may either require higher doses with 3 hour infusion, individual therapeutic drug monitoring or in some cases may not be possible with current ranges of dosing [3–5]. Modeling of levofloxacin in children suggests it may be difficult to achieve Gram-negative pharmacodynamic targets for elavated MICs. Additionally, when Gram-positive targets such as pneumococcus are modeled optimized dosing of levofloxacin achieved Gram-positive targets for MICs only up to 1 mcg/ml in children, despite the susceptibly breakpoint of 2 mcg/mL per Clinical Laboratory Standard Institute [6]. Daptomycin dosing in children also requires special pediatric considerations. As discussed in a review by DeRonde et al. the US Food and Drug Administration (FDA) approved dose for daptomycin in children is 7–12 mg/kg/day, which although much higher than the approved adult dosing, only achieved drug concentrations of 579–620 mg*h/L. Despite being generally greater than what has been reported to be needed for bacteriostatic activity (438 mg*h/L; 95%CI 316–550 mg*h/L), these drug levels were not consistently reaching bactericidal activity (666 mg*h/L 95%CI 501–832 mg*h/L) [7–9]. In these last scenarios, the biggest issue is safety of increased doses, as it could potentially put patients at increased risk of toxicity, especially rhabdomyolysis. Therefore it is essential that pediatric ASPs work closely with clinicians to ensure not only the organisms and medication coverage are considered, but also if pharmacodynamics target attainment can be attained safely in a pediatric patient.

In line with ensuring pediatric patients receive optimal antimicrobial dosing, it is recommended that when feasible pharmacists with understanding of different pharmacokinetics throughout the pediatric lifespan manage medications that require therapeutic drug monitoring [10]. Specifically, neonates have increased volume of distribution for most antibiotics as they are hydrophilic, but have very poor renal function at birth, especially if they are born prior to 35 weeks gestation. In contrast, infants, young children, and even adolescents have faster renal clearance than adult patients.

Although common in adult hospitals, one survey by Welsh et al. reported that only about one-third of pediatric hospitals offer pharmacist-collaborative drug therapy for antimicrobials such as vancomycin or aminoglycosides [11]. As in adults these medications may require dose adjustments. Vancomycin dosing in pediatric patients 3 months old and older is empirically dosed at doses of 60–80 mg/kg/day divided in 4 daily doses for infants and children and 3 daily doses for adolescents but with a maximum dose of 3 g/day [12]. The recently updated vancomycin dosing and monitoring guidelines suggest these doses in pediatrics will achieve similar

AUC to adult patients of 400–600 mg*h/L. At Connecticut Children's, in an effort to evaluate if a change to AUC based dosing was needed and to prevent unecessary blood draws in a child, the use of mg/kg based dosing with an absolute maximum dose was evaluated [13]. The analysis demonstrated the need for these higher doses in pediatric patients, but estimated in retrospective evaluation of a year's patients, that infants and children would be more likely to achieve the goal range if begun on 70 mg/kg/day while adolescents were more apt to overshoot. Therefore, if age based dosing is used alone, a dose of 60 mg/kg/day should be considered [13]. Regardless of the initial dosing, the pharmacist should be involved to ensure that patients achieve goals and are not experiencing toxicity.

In the neonate, each hospital has its own aminoglycoside dosing protocol which often provides a range of doses, based on age and frequency from every 24 h to every 48 h, with patient specific factors prematurity, asphyxia at birth, cyanotic heart disease, or have recently received a non-steroidal anti-inflammatory agent (e.g., ibuprofen or indomethacin for patent ductus arteriosus) requiring less frequent dosing than those born at 30 weeks gestation or later. For children, there is debate on traditional vs once daily dosing due to the concern for once daily dosing even when it has achieved an adequate peak, the child often has cleared the medication faster than the expected post-antibiotic effect. A retrospective review of pediatric patients without cystic fibrosis at Texas Children's reported gentamicin dosing of 7 mg/kg in children 8 years through adolescents, 8.5 mg/kg in children 2 through 7 years, and 9.5 mg/kg daily in those 3 months to 2 years has resulted in peak concentrations in the high 20s while more than 70% in all groups having an undetectable trough (i.e., <0.6 mcg/ml) at 14 h. No patient had an undetectable concentration at 8 h. In cases where there are other antibiotics on board, this concern is likely less of an issue, but the clinical debate continues and often in infants and children traditional 2.5 mg/kg/dose every 8 h is employed [14].

Prevention Through Vaccination

Other significant concerns that effect pediatric ASP are managing patients of variable immune status. The ASP needs to work with clinicians to determine vaccination status and incidents of prior vaccine-preventable diseases. Since the introduction of the pneumococcal conjugate vaccines (i.e., PCV7 and PCV13) antibiotic use among infants decreased from 40.7% to 34.6% [15]. Outpatient antibiotics prescriptions per 1000 population among ages 0 to 19 years have been reduced by 16% [16]. When evaluating patients, when they are not fully immunized, concern may exist for pathogens that would normally be covered by the vaccine. For example, it is recommended that pediatric patients who are under or un-vaccinated receive more broad-spectrum antibiotic therapy with ceftriaxone versus the typical ampicillin to ensure optimal coverage against drug-resistant pneumococcus and *Haemophilus influenzae* if they are admitted for community acquired pneumonia (CAP) [17]. Even more important than recognizing specific conditions for which the patient may need more coverage is the impact that having certain vaccine-preventable infections has on the patient's immune system and susceptibility to disease. Data in recent years have shown that patients that have measles disease are at an elevated risk of dying from an infectious disease for the few years after recovering from measles disease. This increased risk is due to measles disease causing immunity amnesia to non-measles organisms [18].

Pediatric-Specific Antibiotic Considerations

Another important consideration in pediatric stewardship is evaluation of risk versus benefits for medications that are commonly used in adults that may have additional adverse effects possible in pediatric patients [19]. These include medications such as fluoroquinolones and tetracyclines. For the fluoroquinolones, in the initial studies in Beagle puppies an increased risk was found for destruction of the weight bearing joints. More recently, other risks including the promotion of antimicrobial resistance and cardiac toxicity in adults have surfaced. These all come into discussion surrounding their usage in pediatric patients. The American Academy of Pediatrics recommendations state the fluoroquinolone usage in pediatric patients should be limited to treatment of multidrug resistant organisms or oral therapy when the only other option would be intravenous therapy [19]. Tetracyclines as a class have traditionally been avoided in children due to the concern for staining of unerupted teeth. Doxycycline has however been used routinely as the drug of choice for the treatment of Rocky Mountain spotted fever in patients of any age. Multiple studies have been performed showing that for short-term usage (most commonly 7 to 14 days) no differences were seen in the teeth of children who received or did not receive doxycycline therapy. Based upon these studies the American Academy of Pediatrics had updated its guidance in the 2018 RedBook to recommend doxycycline for not only Rocky Mountain spotted fever treatment but also routinely for disease states such as Lyme Disease or skin infections, where there are other options. In their recommendation they state that short courses (up to 21 days of therapy) are unlikely to result in dental staining from doxycycline [20].

Protocol Guided Therapy

As recommended in the Centers for Disease Control and Prevention's (CDC) Core Elements of Hospital Antimicrobial Stewardship, hospitals that care for pediatric patients should have syndrome-specific recommendations and interventions [10]. To accomplish this the ASP should play an integral role in infection related pediatric pathways and protocols. The two most well-researched areas for protocols and impact in pediatric patients include infant sepsis rule-outs and pediatric CAP.

The management of neonatal sepsis varies by age: in the first 7 days of post-natal life, the baby is at risk for early onset sepsis while days 8 through 1 to 2 months is late onset sepsis. For early onset sepsis, a risk calculator developed by Kaiser Permanente (available at https://neonatalsepsiscalculator.kaiserpermanente.org/)

has been shown to reduce unnecessary antibiotic usage in babies who are born at least 34 weeks gestational age. This calculator incorporates the local rate of early onset sepsis in the institution with clinical factors (e.g., gestational age, maternal temperature, duration of rupture of membranes, the mother's group B streptococcus status and if the baby has received antibiotics). The implementation of this calculator has been used at institutions beyond Kaiser and has resulted in decreased admission to neonatal intensive care units and antibiotic exposures, without unanticipated negative outcomes [21, 22]. ASPs have also been shown to decrease overall antibiotic exposure and late-onset sepsis antibiotic use [23].

CAP in pediatric patients is often viral in origin with respiratory syncytial virus one of the most prevalent organisms, especially in those less than 4 years [17]. When vaccinated children do have bacterial CAP, it is most likely caused by pneumococcus and either amoxicillin or intravenous ampicillin is the therapy of choice [17]. ASP programs in conjunction with clinical practice protocols have made significant progress in pediatric hospitals in narrowing CAP therapies. In 2012, Newman and colleagues reported the combination approach was able to modify the most likely antibiotic ordered from ceftriaxone (72% pre-protocol) to ampicillin (63% post-protocol, $p < 0.001$) without any increase in treatment failure [24]. Narrowing therapy to ampicillin (or amoxicillin for outpatients) is significant progress, but it is also important to assess if therapy is needed. Procalcitonin, a marker that is elevated in bacterial infections, has been shown to further impact antibiotic exposures and in certain clinical scenarios may be more specific than other inflammatory markers such as C-reactive protein. Recently, procalcitonin values of less than 0.25 ng/ml have been shown to have a high negative predictive value for bacterial pneumonia [25]. It is essential that ASP or ASP-associated protocols with education are utilized to direct clinicians to discontinue therapy appropriately for low procalcitonin. A recent study in a pediatric medical center showed that procalcitonin alone, without ASP oversight did not result in more patient initiations or stopping of antibiotics within 72 h [26].

Inpatient Stewardship Metrics

Tracking of antimicrobial use is also important for pediatric ASP in order to assess relative and absolute performance in terms of antibiotic utilization. Overall use, as well as usage specific to certain drug classes or disease states can help identify areas of improvement or worrisome trends in prescribing practices. It is important that pediatric ASPs should compare themselves to other pediatric hospitals via collaborations and/or through the CDC's National Healthcare Safety Network (NHSN) antimicrobial use reporting. Pediatric wards are categorized separately than adult wards due to different antibiotics that are used in children, distinct recommendations, and alternative Standardized Antimicrobial Administration Ratio (SAAR) categories as well as data analyzed separately for pediatric wards vs adult wards. The two largest differences between adult and pediatric SAAR categories are fluoroquinolones classification and the specific pediatric SAAR for azithromycin.

Fluoroquinolones are most often used in pediatric patients for nosocomial infections, including treatment of infections caused by *Enterobacter sp.* or *Pseudomonas sp.* (in adults they are categorized in the broad-spectrum community acquired antibiotics). There is limited data to suggest a role for azithromycin as an antibiotic for pediatric respiratory infections. It is no longer included as a recommended agent for acute otitis media due to its limited efficacy against both *Streptococcus pneumoniae* and non-typeable *H. influenzae* [27]. Azithromycin also has not improved outcomes in pediatric CAP. Specifically, in the etiology of pneumonia in the community (EPIC) study, there was no difference in pediatric hospitalization duration in those that received a beta-lactam monotherapy or in combination with azithromycin [28]. As such interpretation of these comparison data should be done by experts in the field of pediatric antimicrobial stewardship, ideally with knowledge of local factors that may impact, or be impacted by, antibiotic utilization.

For example, patients with cystic fibrosis (who have increased inflammation secondary to the chronic infections due to dehydration of the airway surfaces, thick mucus, and lower oxygen concentrations) often require azithromycin for its anti-inflammatory properties to decrease hospitalization if they have chronic colonization with Pseudomonas species, and patients living with sickle cell disease have lower rates of readmission if acute chest syndrome is treated with a cephalosporin plus a macrolide [29]. If such patients were admitted to a single floor, it may report higher than expected usage of azithromycin, but upon review this use would be deemed appropriate.

Outpatient Stewardship Metrics

Outpatient stewardship is another important consideration in pediatric patients. This includes pediatric outpatient offices, urgent care, and emergency department visits. In a recent American Academy of Pediatrics position statement, the academy summarized specific interventions that have been effective in this arena. They include implementing decision support into the electronic prescribing system, providing education to empower providers to not prescribe antimicrobials when they may not be effective or needed, and to have ASP programs review prescribing and provide clinicians with appropriate feedback [30]. It has been our experience that the build of the pathways and order sets paired with education are very important and will make review and feedback easier. Utilizing these ordersets and pathways, the majority of focus can be on individual unique considerations. Additionally, in the outpatient scenario we have found it is harder to have therapies changed once prescribed so ensuring optimal initial prescribing is important. Lastly, just like any other area of practice, keeping up with the literature is essential and incorporating changes when appropriate. Similar to adult patients, recent evidence for multiple types of infections has shown shorter durations of treatment are equally efficacious to longer courses of therapy. Than is traditionally used for many infections is optimal. The most recent of these include 5–7 days for outpatient skin and soft tissue infections and 5 days for community-acquired pneumonia [31, 32].

References

1. Chua K, Fischer MA, Linder JA. Appropriateness of outpatient antibiotic prescribing among privately insured US patients: ICD-10-CM based cross sectional study. BMJ. 2019;364:k5092.
2. Tribble A, Lee B, Handy L, et al. Appropriateness of antibiotic prescribing in US children's hospitals: a national point prevalence survey. Open Forum Infect Dis. 2017;4:S497–8.
3. Cies JJ, Moore WS 2nd, Enache A, et al. B-lactam therapeutic drug management in the PICU. Crit Care Med. 2018;46:272–9.
4. Courter JD, Kuti JL, Nicolau DP. Optimizing bactericidal exposure for beta-lactams using prolonged and continuous infusions in the pediatric population. Pediatr Blood Cancer. 2009;53:379–85.
5. Ellis JM, Kuti JL, Nicolau DP. Use of Monte Carlo simulation to assess pharmacodynamics of beta-lactams against Pseudomonas aeruginosa infections in children: a report from the OPTAMA program. Clin Therapeut. 2005;27:1820–30.
6. Courter JD, Nichols KR, Kazazian C, et al. Pharmacodynamically guided levofloxacin dosing for pediatric community-acquired pneumonia. J Pediatr Infect Dis Soc. 2017;6:118–22.
7. Abdel-Rahman SM, Chandorkar G, Akins RL, et al. Single-dose pharmacokinetics and tolerability of daptomycin 8 to 10 mg/kg in children aged 2 to 6 years with suspected or proved gram-positive infections. Pediatr Infect Dis J. 2011 Aug;30:712–4.
8. Bradley JS, Benziger D, Bokesch P, Jacobs R. Single-dose pharmacokinetics of daptomycin in pediatric patients 3–24 months of age. Pediatr Infect Dis J. 2014;33:936–9.
9. DeRonde KJ, Girotto JE, Nicolau DP. Management of pediatric acute hematogenous osteomyelitis, Part II: a focus on methicillin-resistant Staphylococcus aureus, current and emerging therapies. Pharmacotherapy. 2018;38:1021–37.
10. Centers for Disease Control and Prevention. Core elements of hospital antibiotic stewardship programs. Atlanta, GA: US Department of Health and Human Services, CDC. 2019. https://www.cdc.gov/antibiotic-use/core-elements/hospital.html
11. Welsh C, Miah R, Girotto J. Survey evaluating the practice of children's hospitals having pharmacist collaborative drug therapy management protocols. J Pediatr Pharmacol. 2016;21:494–501.
12. Rybak MJ, Le J, Lodise TP, et al. Therapeutic monitoring of vancomycin for serious methicillin-resistant Staphylococcus aureus infections: a revised consensus guideline and review by the American Society of Health-system Pharmacists, the Infectious Diseases Society of America, the Pediatric Infectious Diseases Society, and the Society of Infectious Diseases Pharmacists. Clin Infect Dis. 2020 Sep 12;71:1361–4.
13. Al-Mazraawy BO, Girotto JE. Comparing vancomycin area under the curve with a pharmacist protocol that incorporates trough and maximum doses at a children's hospital. J Pediatr Pharmacol Ther. 2021;26:740–5.
14. McDade EJ, Wagner JL, Moffett BS, Palazzi DL. Once-daily gentamicin dosing in pediatric patients without cystic fibrosis. Pharmacotherapy. 2010;30(3):248–53.
15. Kinlaw AC, Stürmer T, Lund JL, et al. Trends in antibiotic use by birth season and birth year. Pediatrics. 2017 Sep;140:e20170441.
16. Pediatric ASP toolkit: vaccines to reduce antibiotic use. https://pids.org/pediatric-asp-toolkit/outpatient-settings/vaccines-to-reduce-antibiotic-use/. Accessed 24 March 2021.
17. Bradley JS, Byington CL, Shah SS, et al. The management of community-acquired pneumonia (CAP) in infants and children older than 3 months of age: clinical practice guidelines by the Pediatric Infectious Diseases Society (PIDS) and the Infectious Diseases Society of America (IDSA). Clin Infect Dis. 2011;53:e25–76.
18. Mina MJ, Metcalf CJE, de Stewart RL, et al. Long-term measles-induced immunomodulation increases overall childhood infectious disease mortality. Science. 2015;348:694–9.
19. Jackson MA, Schutze GE, Committee on Infectious Diseases. The use of systemic and topical fluoroquinolones. Pediatrics. 2016;138:e20162706.

20. American Academy of Pediatrics. Committee on Infectious Diseases. Red book: report of the committee on infectious diseases. Elk Grove Village, IL: American Academy of Pediatrics; 2018.
21. Beavers JB, Bai S, Perry J, et al. Implementation and evaluation of the early onset sepsis risk calculator in a high risk university nursery. Clin Pediatr (Phil). 2018;57:1080–5.
22. Warren S, Garcia M, Hankins C. Impact of neonatal early onset sepsis calculator on antibiotic use within two tertiary healthcare centers. J Perinatol. 2017;37:394–7.
23. Nzegwu NI, Rychalsky NR, Nallu LA, et al. Implementation of an antimicrobial stewardship program in a neonatal intensive care unit. Infect Cont Hosp Epidemiol. 2017;10:1137–43.
24. Newman RE, Hedican EB, Herigon JC, et al. Impact of a guideline on management of children hospitalized with community acquired pneumonia. Pediatrics. 2012;129:e597–604.
25. Stockmann C, Ampofo K, Killpack J, et al. Procalcitonin accurately identifies hospitalized children with low risk of bacterial community acquired pneumonia. J Pediatr Infect Dis Soc. 2018;7:46–53.
26. Ross RK, Keele L, Kubis S, et al. Effect of the procalcitonin assay on antibiotic use in critically ill children. J Pediatr Infect Dis Soc. 2018;7:E43–6.
27. Lieberthal AS, Carroll AE, Chonmaitree T, et al. The diagnosis and management of acute otitis media. Pediatrics. 2013 Mar;131:e964–99.
28. Williams DJ, Edwards KM, Self WH, et al. Effectiveness of β-lactam monotherapy vs macrolide combination therapy for children hospitalized with pneumonia. JAMA Pediatr. 2017;171:1184–91.
29. Bundy DG, Richardson TE, Hall M, et al. Association of guideline-adherent antibiotic treatment with readmission of children with sickle cell disease hospitalized with acute chest syndrome. JAMA Pediatr. 2017 Nov;1(171):1090–9.
30. Gerber JS, Jackson MA, Tamma PD, et al. AAP Committee on Infectious Diseases and Pediatric Infectious Diseases Society. Policy statement: antibiotic stewardship in pediatrics. J Pediatr Infect Dis Soc. 2021;10(5):641–9.
31. Pernica JM, Harman S, Kam AJ, et al. Short-course antimicrobial therapy for pediatric community-acquired pneumonia: the SAFER randomized clinical trial. JAMA Pediatr. 2021 May 1;175(5):475–82.
32. Schuler CL, Courter JD, Connely SE, et al. Decrease duration of antibiotic prescribing for uncomplicated skin and soft tissue infections. Pediatrics. 2016;137(2):e20151223.

Antimicrobial Stewardship in Hospital Networks and Health Care Systems

10

M. Sean Boger and Lisa E. Davidson

Introduction

Health care spending accounts for nearly 18% of the U.S. GDP [1]. At the same time, disease treatment has become increasingly complex, resulting in the need for access to specialty care and advanced technology. Single hospital entities are often at a disadvantage in this competitive and costly environment. To improve access and tackle rising health care costs, hospitals have increasingly become integrated to provide continuous and coordinated care [2]. There have been more than 600 hospital mergers in the USA since 2010. Hospitals can be classified as public or private, nonprofit or for-profit, and/or as part of a hospital network or health care system [3].

A hospital system or **health care system**, as defined by the American Hospital Association (AHA), is a multi-hospital or diversified single hospital system, whereby there are two or more hospitals owned, sponsored, or contract managed by a central organization [4]. A multi-hospital system is comprised of at least two facilities owned, leased, sponsored, or contract managed by a central organization. Systems can include one hospital and several separate specialized facilities, or several hospitals and their related satellite facilities. Diversified single hospitals can be categorized as a system by bringing into membership three or more, and at least 25%, of their owned or leased non-hospital pre- or post-acute health care organizations [5]. Therefore, different types of facilities can comprise a health care system (Fig. 10.1). Independent hospitals are not part of a health care system, but they can be a member of a hospital network, and health care systems can be a member of a hospital network.

M. S. Boger (✉) · L. E. Davidson
Antimicrobial Support Network, Department of Internal Medicine, Atrium Health, Charlotte, NC, USA
e-mail: M.Boger@atriumhealth.org; Lisa.Davidson@atriumhealth.org

S. Doron, M. Campion (eds.), *Antimicrobial Stewardship in Non-Traditional Settings*, https://doi.org/10.1007/978-3-031-21710-4_10

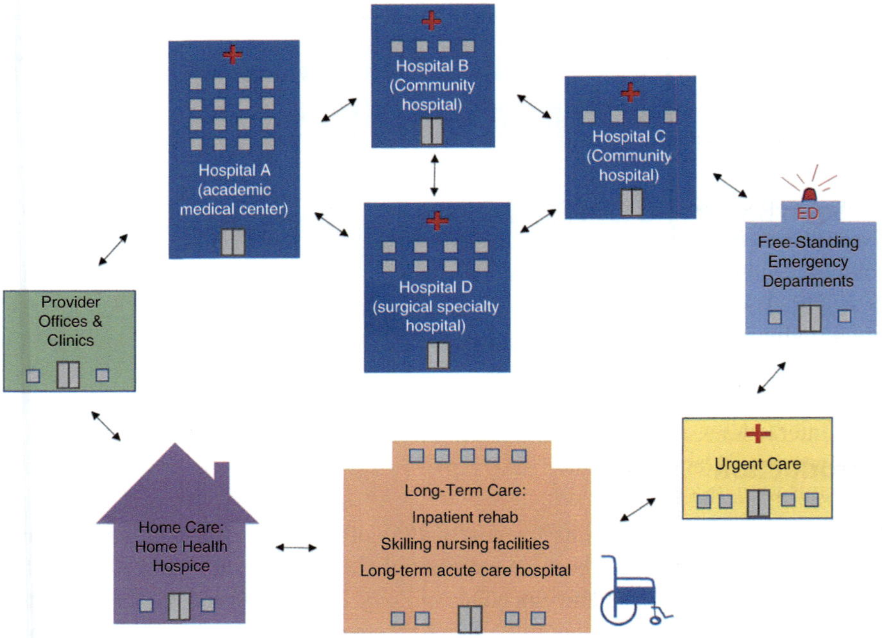

Fig. 10.1 Typical components of a health care system

A **hospital network,** as defined by the AHA, is a group of hospitals, physicians, other providers, insurers, and/or community agencies working together to coordinate and deliver a broad array of health care services to a community [4]. In a hospital network, as opposed to a health care system, the individual facilities are not owned, sponsored, or contract managed by a central organization. Members of a hospital network are administered independently but collaborate, allowing an infrastructure of cross-continuum connectedness and quality transparency [2]. Hospital networks typically have contracts with insurers. Over 1000 new networks were introduced in 2015, with the number of exchange networks totaling over 2900. These networks may be broad, narrow or ultra-narrow, or tiered [6]. With broad networks, more than 70% of all hospitals in a geographic area participate, as opposed to 31–70% participation with narrow networks and 30% or fewer participating in ultra-narrow networks. Tiered network insurance plans group hospitals into tiers based on quality measures and cost in order to steer consumers toward lower-cost hospitals [7]. Narrower networks may limit consumer access to certain providers and increase out-of-network cost of care and, typically, have lower premiums. However, broader networks give consumers more provider choice [6, 8]. An example of a hospital network is Medicare Advantage [8]. The most common form of Medicare Advantage plans is Health Maintenance Organizations (HMOs), which typically require beneficiaries to receive care from in-network providers in order to have the full cost of care covered. Patients participating in Preferred Provider Organizations (PPOs) may receive care from providers outside the network, but their cost-sharing for such care is usually higher compared to that for in-network care. In 2015, Medicare Advantage plans included, on average, about half of the hospitals in a given county in the network.

How does belonging to a hospital system or network affect implementation or participation in an antimicrobial stewardship program (ASP)? In a 2012 AHA survey, more than 70% of hospitals had less than 200 beds, and many did not have ASPs. Studies have shown that small hospitals have similar antibiotic prescribing patterns as large hospitals, including use of broad-spectrum agents [9, 10]. Antimicrobial stewardship can be challenging to implement at smaller community hospitals. It is estimated that nearly one-fifth of the US population resides in rural areas, defined as a county population of less than 50,000. The CDC monitors uptake of hospital ASPs with the National Healthcare Safety Network (NHSN) Annual Hospital Survey. The most recently published data from this survey are from 2015 [11]. 4569 acute care hospitals responded to the 2015 survey. Overall, 48.1% reported uptake of all 7 CDC core elements [12] of antibiotic stewardship. Among hospitals with less than 200 beds, 49.6% met all 7 elements, and only 31.1% of critical access hospitals, defined as rural hospitals with fewer than 25 beds, had an ASP fulfilling all these elements.

The Joint Commission describes improving antimicrobial prescribing practices as a patient safety priority. As a result, a Medication Management (MM.09.01.01) standard for antimicrobial stewardship for hospitals, critical access hospitals, and nursing care centers became effective January 1, 2017 as a requirement for hospitals participating in the Centers for Medicare and Medicaid Services (CMS) program [12]. Elements of Performance for this standard include: leaders establish antimicrobial stewardship as an organization priority; education of staff and licensed independent practitioners about antimicrobial resistance and stewardship practices; an antimicrobial stewardship multidisciplinary team including, when available, infectious disease physician, infection preventionists, pharmacists, and practitioners, with part-time, consultant, or telehealth staff acceptable as members accountable for stewardship activities; drug expertise, use of organization-approved multidisciplinary protocols, collecting, analyzing, and reporting data on the program; and taking action on identified improvement opportunities. Effective January 1, 2020, a new antimicrobial stewardship requirement became applicable to Joint Commission-accredited ambulatory health organizations that routinely prescribe antimicrobials but is not applicable to ambulatory surgery centers or office-based surgery programs [13].

Although having an ASP is a Joint Commission standard, there may be barriers to implementing an ASP in small community hospitals, including the lack of an Infectious Disease (ID) trained physician and access to ID specialty trained pharmacists, minimal dedicated ASP time for physicians and pharmacists, lack of local access to a microbiology laboratory (which can make it harder to form a relationship between the ASP and the lab), and lack of access to a robust EMR with necessary IT support [14]. Small hospitals often lack enough clinical isolates for their own antibiogram. The lack of IT support to assist with obtaining data for tracking and monitoring outcomes may result in more time-consuming, sometimes manual, review of data for ASP activities. Hospital leadership for smaller facilities can be corporate or regional and may not necessarily be at the local level. In these smaller hospitals, if ID-trained pharmacists are not available, alternative leaders of ASP will need to be sought. Opportunities for education and training in antimicrobial stewardship principles include the SHEA antimicrobial stewardship (AMS) Training Course, CDC, or MAD-ID AMS Training Programs.

Advantages for antimicrobial stewardship across health care systems and networks include leveraging information technology, sharing resources including access to larger microbiology laboratories with more testing capabilities and more robust EMR support, pooling and sharing of data, potential access to ID expertise not otherwise available, and collaboration for quality improvement and research. Challenges for antimicrobial stewardship across health care systems include the longer time needed to communicate changes across multiple facilities, identifying accountable persons, engaging medical staff at remote facilities, and the time needed to implement new processes. For these reasons, interventions need to be tailored to the needs and tools available at each facility. If the hospitals in a system have different EMRs or antimicrobial stewardship platforms, it may be difficult to implement uniform care processes. Examples of the system-wide ASP interventions discussed in this chapter are summarized in Table 10.1. This chapter will provide a review of current hospital systems and networks which have successfully implemented large-scale ASP programs.

Table 10.1 Examples of system-wide antimicrobial stewardship interventions

Interventions
Antimicrobial prescribing dashboard with benchmarking and feedback to providers
Traditional facility-based on-site antimicrobial stewardship program (ASP)
Centralized ASP model: on-site antimicrobial stewardship at centralized facilities and virtual antimicrobial stewardship with periodic on-site handshake stewardship for remote facilities
Telehealth ASP for remote hospitals lacking their own ASP
Prospective audit and feedback of antimicrobials
Prior approval of antimicrobials
Evidence-based clinical pathways, guidelines, and order sets • Treatment guidelines for common disease states (e.g., urinary tract infections, community acquired pneumonia) • Guidelines for duration of therapy for common disease states • Guidelines for outpatient parenteral antimicrobial therapy (OPAT) • IV to PO conversion protocols • Renal dosing guidelines for antimicrobials • Vancomycin and aminoglycoside dosing and monitoring
Establishing indications for antimicrobials
Clinical decision support with standardized rules, justification alerts
48–72-h antibiotic "timeout"
Educational newsletters for providers
System-wide educational campaigns

Atrium Health

Atrium Health is a large integrated health care system in the southeastern United States. The system (primary enterprise, regional enterprises, and affiliated enterprises) includes over forty hospitals in North Carolina, South Carolina, and Georgia. The primary enterprise facilities of Atrium Health (formerly Carolinas HealthCare System) are composed of 11 hospitals including an 872-bed teaching hospital, a 214-bed children's hospital, four rehabilitation facilities, two behavioral health facilities, two inpatient hospice facilities, over 30 urgent care and free-standing emergency departments, and 5 skilled nursing facilities. In 2019, Atrium Health developed a strategic combination with Navicent Health to expand health care services in central and southern Georgia. Navicent Health includes a tertiary teaching hospital and a dedicated pediatric hospital, in addition to a rehabilitation facility and community hospitals.

The Atrium Health Antimicrobial Support Network (ASN) was conceived in 2013 and is comprised of four teams (three adult and one pediatric) of ASN pharmacists grouped geographically. The goal is to provide standardized ASP services to all facilities 7 days per week. Program development and education as part of the structuring of this network was provided by a centralized leadership and operational team. When initially implemented at six hospitals in 2013, local facility-based non-ID trained pharmacists performed daily stewardship activity [15]. A gap analysis showed that a local staffing model only covered about half of the patients eligible for review. A centralized network model was then implemented. Implementation of the centralized ASP model was more effective at reducing antimicrobial use than the initial traditional facility-based approach: the shift to a centralized model resulted in an increase in interventions from 3897 to 11,612, covering more than 95% of all eligible patients. The impact of the ASN on three representative facilities in the Atrium Health system (a large teaching hospital, a large community hospital, and a small community hospital) is illustrated in Fig. 10.2. Examples of baseline interventions for these 3 illustrative facilities during the first year of the ASN program are demonstrated in Fig. 10.3.

The ASN teams use a combination of virtual and on-site daily review. These teams are overseen by a centralized physician medical director and pharmacy manager. A pediatric division and an ambulatory division were also created. System-wide campaigns have included diagnosis and management of *Staphylococcus aureus* bacteremia, asymptomatic bacteriuria (ASB), and *Clostridium difficile* (*C. difficile*) testing and treatment [16, 17]. The ASN from early adoption of the CDC Core Elements initiative saw a 25% reduction in overall antibiotic prescribing over five years. Examples of disease state-specific interventions implemented at the flagship teaching hospital in the Atrium Health system and their associated outcomes are shown in Table 10.2 [18–20].

Atrium Health is both a hospital system and an integrated health care network. Integration outside of the hospital allows for ASP expansion beyond the traditional inpatient setting. The Carolinas HealthCare Outpatient Antimicrobial Stewardship Empowerment Network (CHOSEN) is a collaboration with more than 150

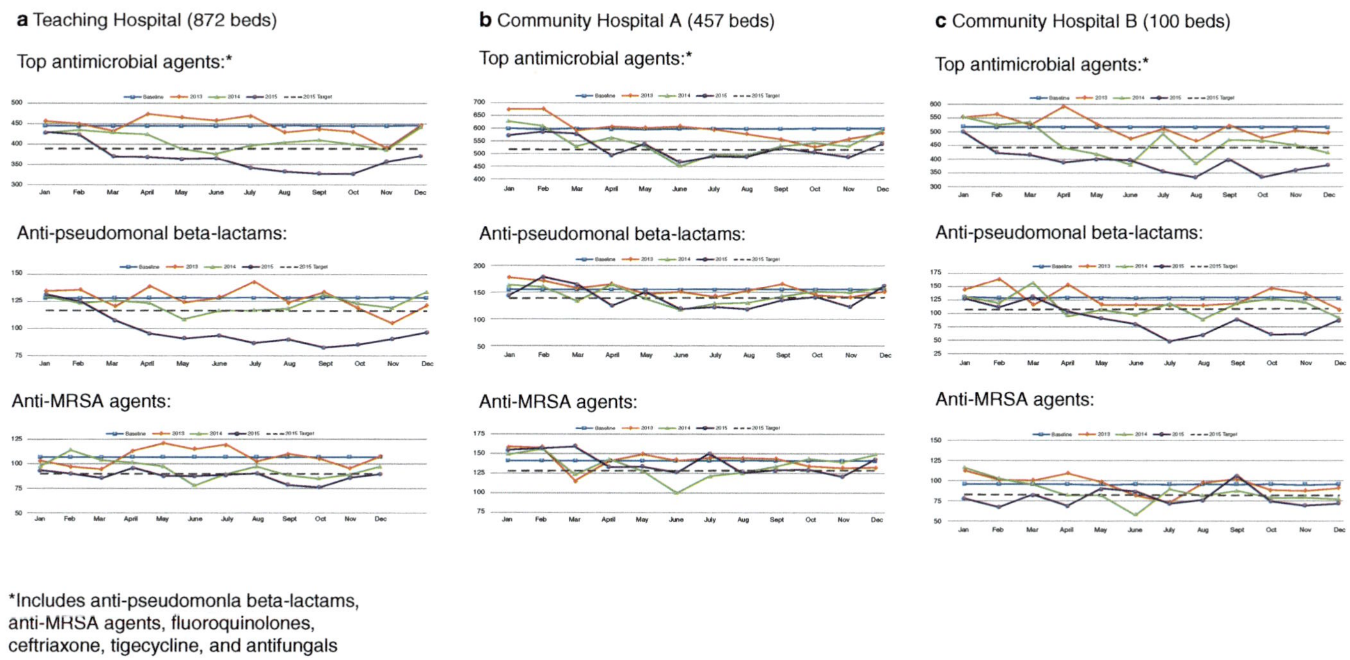

Fig. 10.2 Impact of system-wide antimicrobial stewardship program implementation on antimicrobial usage using three representative facilities in the Atrium Health system (**a**=large teaching hospital, **b**=large community hospital, and **c**=small community hospital). Data are in days of therapy (DOT) per 1000 patient-days from baseline and the first 3 years of the program

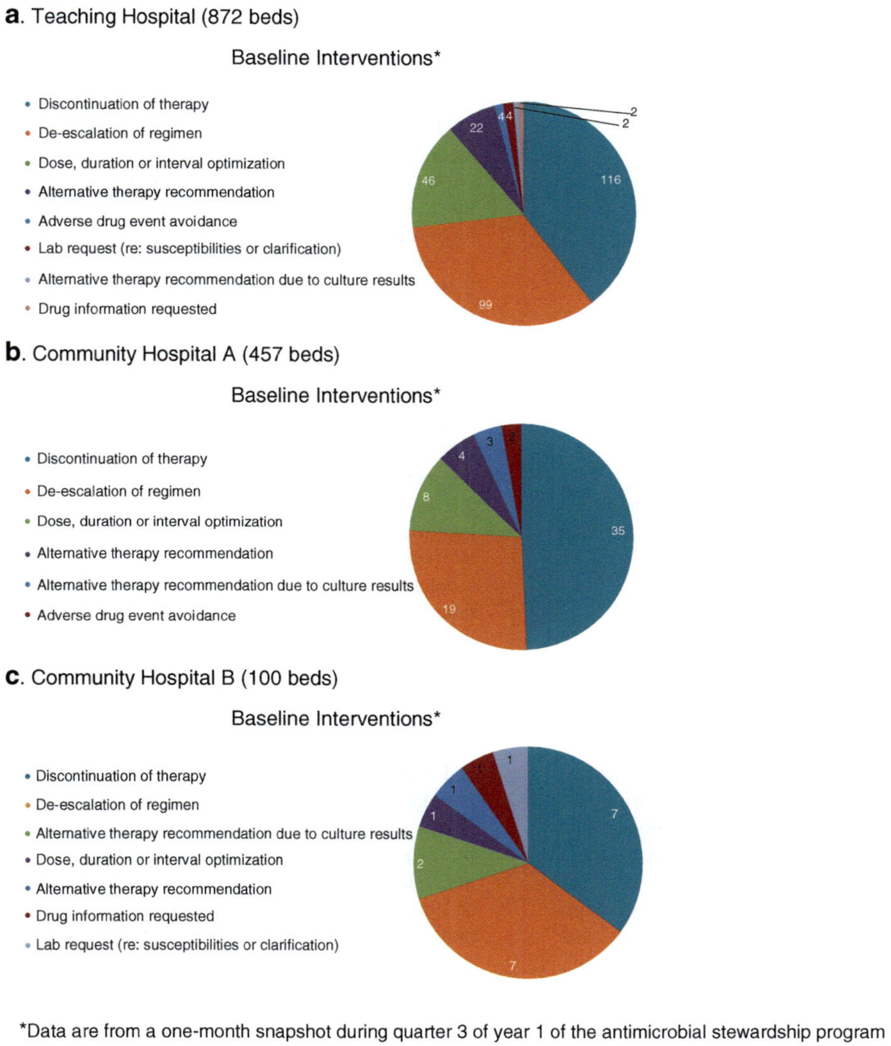

Fig. 10.3 Baseline interventions during a one-month snapshot of the first year of the antimicrobial stewardship program using three representative facilities in the Atrium Health system (**a**=large teaching hospital, **b**=large community hospital, and **c**=small community hospital)

ambulatory practices across Atrium Health [21]. The aim of this collaborative was a 20% reduction in outpatient antimicrobial prescribing over 2 years with a focus on acute respiratory conditions. Resources including provider and patient education were introduced with a consumer webpage and via the provider intranet. Focused provider education was provided for ambulatory practice specialties of Urgent Care, Family Medicine, Internal Medicine, and Pediatrics [22]. An innovative antibiotic prescribing dashboard for agents used for respiratory tract infections was developed in Microsoft Power BI and launched in March 2018, with means to view prescribing

Table 10.2 Examples of disease state-specific interventions implemented at a large teaching hospital in the Atrium Health system

Intervention	Goals	Outcomes
Antiretroviral therapy (ART) review • All inpatients on ART are reviewed daily	• Reduce ART-related medication errors • Increase linkage to care for persons living with HIV	• Proportion of patients receiving ART with medication-related errors decreased from 70% to 25.7% • Proportion of patients with linkage to care increased from 19% to 39.6%
Immunocompromised ID rounding service • ASN pharmacists began rounding with solid organ transplant (SOT) and hematopoietic stem cell transplant (HSCT) ID teams on alternating days with daily chart review	• Decrease antimicrobial utilization in SOT and HSCT programs by 5% • Increase intervention acceptance rates by 10%	• Anti-pseudomonal agents decreased by 9%, anti-MRSA agents decreased by 22%, and antifungals decreased by 5% • Intervention acceptance rate increased from 71% to 83% • *C. difficile* infection rates for HSCT patient care unit decreased
Handshake stewardship with prospective audit and feedback with hepato-pancreato-biliary surgical services (HPBSS) • Twice-weekly rounding with HPBSS with ASN PharmD • Routine remote antimicrobial stewardship on non-rounding days • Treatment guidance documents developed for common disease states (e.g., liver abscesses, necrotizing pancreatitis)	• Decrease antibiotic utilization of carbapenems and anti-pseudomonal agents • Increase ASN intervention acceptance rates	• Utilization of carbapenems decreased by 62.3% and anti-pseudomonal agents decreased by 18.6% • ASN intervention acceptance rate increased by 31% • Reduced rates of carbapenem resistant Enterobacteriaceae (CRE) hospital associated (HA)-infection/colonization and HA-CDI were observed

ASN, antimicrobial stewardship network; HIV, human immunodeficiency virus

data by indication and antibiotics, comparing year-to-year and rolling 24 months with the option to drill down to practice and provider level data [23].

There are many advantages of sharing resources between facilities in this model of a local regional health care system. A unified electronic health care record (EHR) and antimicrobial stewardship surveillance system allows for uniform data collection and comparisons between facilities. Smaller facilities that, for geographical reasons, do not have daily on-site ID pharmacists have access to an identified team ASN pharmacist physically located at one of the larger facilities who is able to perform daily stewardship review remotely. Back-up to pharmacists is readily available by one of the ASP medical directors who are located at one of the larger centralized facilities. The ASP pharmacists serving the smaller hospitals regularly go onsite to assess the needs of the individual facilities and to provide education. Since this is a large health care system, there is the ability to evaluate how outcomes differ when tailored to the diverse individual facilities and the ability to pool data for larger outcomes. All the facilities have access to the centralized microbiology laboratory and tailored antibiograms published according to the isolates at each facility. A unified system pharmacy and therapeutics (P and T) committee provides access to review by content experts who might otherwise not be readily available and representation for these facilities. New antimicrobials approved at the system P and T level are available for use with the same indications and restrictions at the smaller regional facilities as for the larger facilities. All the facilities share access to evidence-based clinical pathways, guidelines, and order sets, which are implemented in a standardized fashion across the facilities.

Ascension Health

Ascension Health, based in St. Louis, MO, is the largest non-governmental, non-profit hospital system in the USA, with over 145 hospitals and 2600 sites of care in 19 states and the District of Columbia [24]. It is currently the largest Catholic health system in the world. In 2015, a Center of Excellence was created to focus antimicrobial stewardship efforts system wide. This led to the establishment of local ASPs in 89 hospitals. Each hospital ASP team chose interventions based on local needs, available resources and expertise, with intervention recommendations grouped into five categories: prospective monitoring of broad-spectrum gram-negative agents, prospective monitoring of anti-MRSA agents, order set standardization for treatment of common conditions, order set standardization of surgical antimicrobial prophylaxis, and establishing appropriate indications for all antimicrobials. Results were compared based on hospital bed size (small: <100 beds, medium: 100–300 beds, and large: >300 beds). Small hospitals focused more on prospective monitoring of gram-negative and anti-MRSA agents, medium-sized facilities concentrated on order set standardization, and large hospitals more often focused on antibiotics used for surgical prophylaxis. The majority (85%) of Ascension Health hospitals focused on prospective monitoring of broad-spectrum anti-gram-negative agents, nearly two-thirds focused on prospective monitoring of anti-MRSA agents, and 30% focused on order set standardization for treatment of common infectious disease conditions as a local initiative [25].

System-wide results of the ASP led to a 15.9% reduction in hospital-onset *C. difficile* infections and more than a 12% reduction in systemic antibiotic use for fiscal years 2016 and 2017, with a 40% reduction in fluoroquinolone use alone [26]. There was an annualized $2.4 million in savings for hospitals in the Ascension network due to a reduction in systemic antibiotic use from FY15 to FY16 after each hospital developed physician-pharmacist co-led antimicrobial stewardship teams to ensure compliance with evidence-based antimicrobial use, culture orders, and standardized disease state management processes [27]. Antimicrobial use over time, measured by defined daily dose (DDD) per 1000 patient-days, was shared with each facility monthly to identify successes and opportunities so that a plan of action could be developed for each site. Hospitals were also directly engaged by phone calls and site visits. A dashboard was developed to demonstrate the successes and opportunities for improvement at each facility with regard to the use of daptomycin, linezolid, tigecycline, cephalosporins, carbapenems, quinolones, and aztreonam [28]. After this dashboard was instituted, use of these agents decreased by 2–43%.

In a study of 69 hospitals in the Ascension network from 2010–2012, smaller size hospitals had more antimicrobial use than larger facilities, and those with a lower case mix index had greater antimicrobial use and cost as compared to hospitals with a higher case mix index. This highlights the importance of conducting antimicrobial stewardship across the system [29]. In 2015, a clinical decision support system (CDSS) was implemented with standardized rules at 57 hospitals in the system [30]. These stewardship rule algorithms interfaced between laboratory, microbiology, medication profile, surgery, and radiology reports and included rules addressing bug-drug mismatches, broad-spectrum and niche antibiotic agent reviews at 72 h, methicillin-sensitive *S. aureus* antibiotic streamlining, and penicillin allergy assessment. The goal of this implementation was to improve pharmacist workflow to allow therapeutic interventions to be made in real time 24 h a day. The CDSS resulted in 74,232 interventions in 1 year, during which time there was a 10% reduction in total antibiotic use.

Intermountain Health Care

Intermountain Health Care, based in Salt Lake City, UT, is a non-for-profit system of 24 hospitals and 215 clinics, with nearly 2400 employed physicians and advanced practitioners [31]. It is the largest healthcare provider in the Intermountain West. Intermountain Healthcare includes both large academic medical centers and small critical access hospitals that are geographically widespread. Intermountain Health does have the advantage of a common electronic medical record. A novel telehealth ASP was developed, whereby an ID physician and ID pharmacist located at the centralized facility for the system provide data, mentoring, and consultation remotely via a toll-free line. The small hospitals in the system were individually responsible for their own ASP. In 2017, it was estimated that system hospitals with fewer than 70 beds had, on average, less than 5 h per week of dedicated time for ASP activities, as compared to an average of 5–10 h per week for hospitals having 70 to 150 beds [14].

The Stewardship in Community Hospitals Optimizing Outcomes and Resources (SCORE) study, which had a cluster-randomized intervention design, assessed the impact of implementing an ASP at 15 small community hospitals (ranging in size from 14 to 146 licensed beds) which, as of 2016, comprised 25% of Intermountain Health Care beds [32]. These small community hospitals did not have on-site ID physician support and, prior to this study, did not have routine ID consultation available nor did they have access to an ASP. Hospitals were randomized to one of the three types of ASP interventions of escalating intensity and increasing reliance on centralized support from ID physicians and pharmacists. Antibiotic best practices were used at all facilities, including IV to PO conversion, antibiotic indications, and a 48-h antibiotic "timeout." Program 1 hospitals were provided basic antibiotic stewardship education and tools, access to an infectious disease hotline staffed 24 h a day, 7 days a week, by an attending ID physician, and a monthly hospital-specific antibiotic usage dashboard. Program 2 hospitals received the interventions provided to Program 1 facilities along with advanced antibiotic stewardship education, including online learning modules and a half-day live workshop, audit and feedback for a select group of antibiotics, and antibiotic restrictions which were regulated locally. Guidelines for use of these restricted antibiotics, including acceptable uses and recommended monitoring parameters, were provided to these sites. Program 3 hospitals received the interventions given to Program 2 facilities plus audit and feedback on an expanded list of antibiotics, and an ID-trained clinician approved antimicrobial restrictions and reviewed certain microbiology results including positive blood cultures and cultures with highly resistant Enterobacteriaceae. Total antibiotic use declined by 11% from baseline during the intervention period (rate ratio 0.89; CI 0.80–0.99) in program 3 hospitals but remained unchanged in program 1 and 2 hospitals. Use of broad-spectrum antibiotics was reduced by 24% (RR 0.76; CI: 0.63–0.91) in program 3 hospitals but no significant reductions in these agents were seen in program 1 and 2 hospitals. Use of restricted antibiotics declined by 86% in program 3 hospitals as compared to 52% in program 2 hospitals.

Another study from the Intermountain Healthcare system evaluated antibiotic use in 15 small community hospitals (SCHs) versus four large community hospitals (LCHs) [9]. Antibiotic agents were classified into five groups based on spectrum and ability to treat multidrug-resistant organisms, with category 1 being narrow-spectrum and category 5 being the broadest-spectrum or those associated with treatment of multidrug-resistant organisms. Antibiotic use varied greatly among the SCHs and was like that in the LCH, despite SCHs having a significantly lower case mix index. Excluding miscellaneous care units, the proportion of total antibiotic use accounted for by broad-spectrum antibiotics varied significantly among the SCHs, from 8% to 36% of total use. Over the three-year study period, the aggregated antibiotic use rate was stable for the SCHs, with a relative rate difference per year of 1.48% (95% CI: −0.39% to 3.39%), and decreased for the LCHs, with a relative rate difference per year of −3.54% (95% CI: −4.44% to −2.63%). None of the SCHs had active ASPs during the study period, and three of the LCHs had active ID-led ASPs.

In a retrospective cohort study of over 1 million urgent care encounters in the Intermountain Healthcare network, 34% of encounters resulted in antibiotic prescriptions [33]. ICD10 codes were assigned to one of the three tiered groups. Tier 1 codes (e.g., pneumonia and urinary tract infection) were ones where antibiotics are almost always indicated, Tier 2 codes (e.g., sinusitis, acute otitis media) were ones where antibiotics may be indicated, and Tier 3 codes (e.g., bronchitis and nausea) were ones where antibiotics are never indicated. Of respiratory tract infections, antibiotics were prescribed for 64.8% of Tier 2 and 21.3% of Tier 3 encounters.

Pharyngitis, sinusitis, and acute otitis media accounted for more than half of all antibiotic prescriptions. For these three diagnoses, only 69.2% of prescriptions were for a first-line antibiotic. Significant variability was noted at the provider level, ranging from 3% to 94% of encounters for respiratory tract conditions receiving antibiotics. Interestingly, the mean patient satisfaction score was only slightly higher for visits where an antibiotic was prescribed as compared to visits where an antibiotic was not prescribed. In July 2019, antimicrobial stewardship interventions were deployed in urgent care settings [34]. Electronic healthcare record tools were implemented including prompts for delayed prescribing for certain respiratory tract infections and justification alerts for azithromycin. Monthly antibiotic prescribing data were provided and leadership commitment posters were displayed in clinics. Delayed prescriptions increased across the network and respiratory tract antibiotic prescribing began to decline after this intervention.

Kaiser Permanente

Kaiser Permanente (Oakland, CA), the prototype example of an HMO, it is the largest nonprofit integrated health care system in the USA, serving more than 12 million members. There are 39 hospitals, 724 medical offices, and more than 20,000 physicians in 8 states and the District of Columbia [35]. It is comprised of Kaiser Foundation Hospitals and its subsidiaries, Kaiser Foundation Health Plan, Inc., and the Permanente Medical Group. In Northern California, Kaiser Permanente has an estimated 44% of the market share. Because this model of care is prepaid, there is a focus on prevention and wellness [36]. Kaiser Permanente Washington partnered with Choosing Wisely to reduce low value tests and treatments that are usually unnecessary and may cause harm. The proportion of visits for upper respiratory tract infections leading to antibiotic prescribing decreased from 37% to 27%, with a goal of less than a 20% prescription rate for these indications in 2018 [37]. In 2016, 3800 unneeded outpatient antibiotic prescriptions were avoided because of this initiative.

A study by Kaiser Permanente West, Los Angeles Medical Center, evaluated the cost and quality outcomes achieved by implementation of ASPs in a pre-post analysis [38]. Two ASP sites were selected as the study sites, and three similar sites within the health system which were not included in the ASP initiative served as a comparison group. Utilization of 15 targeted antimicrobials (anti-pseudomonal, quinolones and anti-MRSA agents) and their associated costs were compared. Significant reductions in consumption of targeted antimicrobial classes were seen at the ASP sites, with the largest decreases observed in anti-pseudomonal agents (18.3%) and fluoroquinolones (16%). At the study sites, the aggregate antimicrobial cost for targeted agents decreased by 17.3% during the study period relative to the preceding year, whereas cost for these agents increased by 9.1% at the comparator sites. ASPs with a dedicated ID pharmacist showed an aggregate savings of $228,353.

In a stepped-wedge cluster-randomized intervention study of 105 primary care and 21 urgent care clinics in the Kaiser network, clinical decision support (CDS) was implemented along with provider education to determine the impact of these interventions on antibiotic prescribing for acute sinusitis [39]. The CDS was evidence-based, automated in the EMR, and initiated at the time of antibiotic prescribing for an acute sinusitis diagnosis. In 21,949 encounters for acute sinusitis, CDS reduced overall antibiotic use [adjusted OR 0.78 (95% CI 0.71–0.87)] and guideline-concordant antibiotic use increased by 14%. The CDS had a sustained impact on antibiotic prescribing, whereas provider education had a large initial effect that did not persist.

The Veterans Health Administration

The Veterans Health Administration (VHA) is the largest integrated health care system in the USA, with greater than 170 VA Medical Centers and over a thousand outpatient sites of care, treating over 9 million enrolled Veterans [40]. The VHA is one of the largest integrated health care systems reporting data to NHSN. VHA's system-wide antimicrobial stewardship program initiative began in mid-2010 with a dedicated task force and consisted of a series of in-person educational conferences, the creation of online resources, monthly webinars, and sample policies [41]. Examples of policies made available include avoidance of double anaerobic coverage, monitoring of outpatient parenteral antimicrobial therapy, vancomycin de-escalation, broad-spectrum gram-negative de-escalation, pneumonia duration of therapy, and *S. aureus* bacteremia treatment duration. These resources were available to all VHA medical centers for elective use as each site deemed appropriate. The stewardship champions at local facilities were encouraged to take an "a la carte" approach to implementing stewardship practices. The VHA demonstrated its strong commitment to antimicrobial stewardship though the establishment of VHA Directive 1031 which requires all VHA medical facilities to implement, maintain, and annually evaluate ASPs [42].

Implementation of the VHA National Antimicrobial Stewardship Initiative was shown to have many beneficial effects across the VHA system [41]. From 2011 to 2015, the number of hospitals with a defined antimicrobial stewardship team increased from 41% to 89%, and the number having a written policy for stewardship increased from 17% to 92%. From 2011 until 2015, the percentage of hospitals having prior approval procedures for antimicrobials increased from 59% to 87%. During this time frame, the proportion using post-prescription review increased from 8% to 71%. From 2012 until 2015, significantly more facilities reported review for duration of therapy and *C. difficile* interventions, and there were significant reductions in sites reporting challenges with provider buy-in, administrative support, and pharmacy support. A 12% reduction in overall antibiotic use, measured in days of therapy per 1000 bed days of care, was seen from 2008 to 2015. No change was seen in the yearly proportional distribution of specific antimicrobial agents, classes, and groups, which the authors hypothesize reflects that the decrease in antimicrobial use was not related specifically to one agent or one class of agents but, rather, to an overall global decline in usage across antimicrobial classes. Hospital acute-care and intensive-care lengths of stay remained stable, while readmission and mortality rates decreased after implementing the initiative. From 2012 to 2015, there were significant reductions in rates of *C. difficile* infection, which were likely also at least in part the result of a concurrent VHA-led *C. difficile* prevention program [43].

A pilot study assessed the feasibility of telehealth to support antimicrobial stewardship at VA medical centers having limited access to ID-trained specialists, and the outcomes were reported for two facilities [44]. At each intervention site, providers, pharmacists, infection preventionists, nurses, and off-site ID physicians formed a videoconference antimicrobial stewardship team (VAST) that met weekly to discuss cases and to perform antimicrobial stewardship-related education. The implementation varied at the two study facilities—at one site, an infection preventionist championed the VAST and determined which cases to discuss and present, whereas a pharmacist served as the champion at the other site. Leadership support was significant at one site yet limited at the other facility. On average, 3.1–3.5 cases for review were discussed per session. Interviews at the study sites revealed that participants valued the VAST and felt that the educational component was beneficial and improved patient care.

At a 160-bed long-term care facility (LTCF) affiliated with a large urban VA medical center, an LTCF-Infectious Disease (LID) team was established with an Infectious Disease physician and a nurse practitioner examining residents at the LTCF once weekly and available for remote consultation the remainder of the week [45]. Urgent questions were fielded immediately whereas routine questions were seen during weekly rounds. The most common reasons for consultation were *C. difficile* infection and conditions ultimately deemed non-infectious, including asymptomatic bacteriuria and colonized wounds. There was a >95% intervention acceptance rate for the LID consult service. After initiation of the LID consult service, total antimicrobial use decreased by 30%, with the greatest reduction observed in oral agents. A 48% reduction in piperacillin-tazobactam use was noteworthy. The antibiotic use at the affiliated VA hospital without this service did not change during the study.

A study to determine associations between ASPs, facility characteristics, and inpatient antimicrobial utilization was performed utilizing a survey completed by 130 VA facilities with acute care services [46]. For the survey, 34 variables were identified across three domains (evidence, organizational context, and facilitation) and assessed against four dependent measures of antimicrobial utilization (aggregate acute care antimicrobial use, antimicrobial use in patients with non-infectious primary discharge diagnoses, missed opportunities to convert from parenteral to oral therapy, and double anaerobic coverage). Variables in this study associated with favorable changes in antimicrobial utilization included the presence of post-graduate physician or pharmacy training programs at the facility, the number of antimicrobial-specific order sets, the presence of pharmacists and/or ID attendings on acute care ward teams, and formal ID training of the lead ASP pharmacist. Variables associated with unfavorable measures included bed size and the level of engagement with the VA Antimicrobial Stewardship Task Force online resources.

Michigan Antibiotic Resistance Reduction Coalition

The Michigan Antibiotic Resistance Reduction (MARR) Coalition is a multi-stakeholder nonprofit organization supported by an annual grant from the CDC established to improve the use of antimicrobial agents in communities throughout the state of Michigan via the collaborative educational efforts of academic, government, and community partners [47]. The goals of this coalition are to provide the infrastructure and resources including educational interventions to change provider prescribing habits and improve patient communication skills, to reduce antimicrobial resistance in the state of Michigan, to provide a forum for sharing efforts for antimicrobial resistance reduction, and to secure accurate data on antibiotic prescribing trends of health care providers in the outpatient setting and on consumer use of antibiotics. Examples of healthcare provider education provided by the coalition include a long-term care tool kit, CME accredited presentations, and a webcast for nurses.

This collaborative conducted multimodal physician education on best practices for antimicrobial use including enhanced antimicrobial documentation, improved quality and accessibility of local clinical guidelines, and a 72-h antibiotic timeout. Facilitators engaged hospitalists, provided performance feedback, and limited the scope of interventions based on the practice site [48]. There were two academic medical centers, two community teaching hospitals, and one community nonteaching hospital. In two hospitals, there was significant improvement in complete antimicrobial documentation (4% to 51% and 8% to 65%, $p < 0.001$ for each comparison). The institution of antibiotic timeouts resulted in optimization or discontinuation of antimicrobials 30% of the time. A retrospective cohort study evaluating over 11,000 patients admitted with pneumonia or positive urine cultures in 48 hospitals in this collaborative found that prospective audit and feedback or prior approval targeting fluoroquinolones was associated with significantly lower use of this class of agents (37% vs 48%, $p = 0.01$) during hospitalization but not at discharge [49].

In conjunction with the Michigan Society of Health-System Pharmacists, an ID pharmacy networking group was established to explore ways in which to expand antimicrobial stewardship efforts across the state of Michigan. To determine a baseline understanding of the needs and capabilities of stewardship in various health care settings in the state, a 30-question electronic survey was developed to assess stewardship practices and priorities according to the suggested core strategies from the IDSA [50]. Forty-five percent of respondents represented facilities with less than 150 licensed beds and 58% represented sites with fewer than 15 ICU beds. Fewer than half of facilities reported that their ASP team was multidisciplinary in nature.

Vizient

Vizient is a large member-driven, health care performance improvement company designed to drive improvement in health care by connecting members to solutions that boost quality and deliver cost-effective care [51]. Membership includes academic medical centers, community hospitals, pediatric facilities, integrated health delivery networks, and non-acute health care providers. Vizient is ranked as the third largest health care management consulting firm, with more than 550 consultants dedicated to improving members' performance [52]. Vizient is an example of the twenty-nine Practice Transformation Networks (PTN) selected by CMS to help clinicians integrate quality and process improvements. As a PTN, all clinicians can participate in the Vizient collaborative, unless they are already participating in other programs involving practice transformation, such as the Medicare Shared Savings Program, the Pioneer Accountable Care Organizations (ACO) programs, or the CMS Multi-Payer Advanced Primary Care Practice. Areas of focus include reducing unnecessary clinical and operational variation, improving clinical outcomes, ensuring a sustainable cost structure, and executing strategies for market success. The Vizient University Health System Consortium is a large, comprehensive network of academic medical centers in the USA, with more than 90% participation in the consortium. The Vizient Antimicrobial Stewardship Collaborative has focused on reducing harm associated with delivery of care and establishing ASP in hospitals, and participants have worked on projects that best fit their organizations.

A 2016 survey by Vizient and Virginia Commonwealth University developed based on the CDC Core Elements of ASPs calculated a stewardship score for each organization with a total possible score of 20, with high-performing ASPs categorized as having a score of 13–20 and low-performing ASPs having a score of 1–8. Only 34% of 211 hospitals surveyed were following all the CDC Core Elements for ASPs. High-performing ASPs had several defining characteristics: 82% were led by organizations connected to a health system, 89% provided formal infectious disease training to staff, and 76% compensated physicians for participating in ASP. Academic medical centers were more likely to have a high-performing ASP compared to community hospitals. Medium and small community hospitals had the least comprehensive programs, accounting for more than half of low-performing ASPs. Forty-five percent of facilities did not have a formal written statement of leadership support. One-fourth of hospitals surveyed conducted routine antibiotic "timeouts," and only 6% of programs required documentation of indication, dose, and duration for all antibiotics in the patient medical record or during order entry [53].

Future Directions

Hospital collaboratives such as health care systems and networks will likely continue to expand efforts for antimicrobial stewardship to other health care settings, such as nursing homes and other long-term care facilities, rehabilitation facilities, the outpatient setting such as clinics and urgent care, and in the emergency department. Expanding telehealth capabilities into the realm of antimicrobial stewardship will likely be another focus. These activities will help to further coordinate and improve antimicrobial stewardship across the continuum of care both in local communities and across regions. This would allow antimicrobial stewardship to assist with transitions of care from the inpatient to outpatient setting and vice versa.

Conclusion

Health care systems and hospital networks allow individual facilities to collaborate and share access to resources, clinical expertise, and robust data to improve quality of care across the continuum, whether they intersect with smaller or larger facilities. This allows hospitals, particularly smaller facilities, to better control the cost of care while also sharing resources for optimizing quality and patient safety. Advantages of participation in health care systems or hospital networks for antimicrobial stewardship include access to ID-trained pharmacists and physicians for expertise and educational opportunities, leveraging support to improve the EMR and antimicrobial surveillance systems to optimize daily ASP activities, and the ability to collaborate to improve outcomes. These systems and networks may enable small to medium-sized facilities to meet the CDC and Joint Commission requirements for antimicrobial stewardship when they might otherwise not be able to do so from a resource standpoint. It is anticipated that trends toward more integration in health care will continue to increase as accountability for health care outcomes spreads across the entire care continuum. Therefore, these antimicrobial stewardship opportunities are poised to expand the horizon across all types of healthcare settings.

References

1. Hartman M, Martin AB, Espinosa N, et al. The National Health Expenditure Accounts Team. National health care spending in 2016: spending and enrollment growth slow after initial coverage expansions. Health Aff. 2018;37:150–60.
2. Bush J. The future of the hospital is the network. Becker's Hospital Review April 2016. https://www.beckershospitalreview.com/the-future-of-the-hospital-is-the-network.html. Accessed 4 May 2019.
3. Buckel WR, Veilette JJ, Vento TJ, et al. Antimicrobial stewardship in community hospitals. Med Clin North Am. 2018;102:913–28.
4. AHA Resource Center. Fast Facts on U.S. Hospitals. In: AHA Hospital Statistics. American Hospital Association. 2017 ed. https://www.aha.org/statistics/fast-facts-us-hospitals-2017. Accessed 10 July 2019.
5. Kovner AR. Hospitals, Chapter 6. In: Kovner AR, Jonas S, editors. Health care delivery in the United States. 7th ed. New York: Springer Publishing Company; 2002. p. 145–72.
6. McKinsey Center for U.S. Health System Reform. Hospital networks: perspective from four years of the individual market exchanges. May 2017. McKinsey and Company. https://healthcare.mckinsey.com/sites/default/files/2017%20Hospital%20networks%20-%20Perspectives%20from%20four%20years%20of%20the%20individual_%20exchanges%20vF.pdf. Accessed 4 May 2019.

7. Frank MB, Hsu J, Landrum MB, et al. The impact of a tiered network on hospital choice. Health Serv Res. 2015;50:1628–48.
8. Jacobson G, Trilling A, and Neuman T. Medicare advantage hospital networks: how much do they vary? Kaiser Family Foundation Report. June 2016. http://files.kff.org/attachment/Report-Medicare-Advantage-Hospital-Networks-How-Much-Do-They-Vary. Accessed 6 May 2019.
9. Stenehjem E, Hersh AL, Sheng X, et al. Antibiotic use in small community hospitals. Clin Infect Dis. 2016;63:1273–80.
10. Magill SS, Edwards JR, Beldavs ZG, et al. Prevalence of antimicrobial use in US acute care hospitals, May-September 2011. JAMA. 2014;312:1438–46.
11. O'Leary EN, van Santen KL, Webb AK, et al. Uptake of antibiotic stewardship programs in US acute care hospitals: findings from the 2015 National Healthcare Safety Network Annual Hospital Survey. Clin Infect Dis. 2017;65:1748–50.
12. Joint Commission New Antimicrobial Stewardship Standard. https://www.jointcommission.org/assets/1/6/New_Antimicrobial_Stewardship_Standard.pdf. Accessed 4 June 2021.
13. Joint Commission Ambulatory Antimicrobial Stewardship Standard. https://www.jointcommission.org/-/media/tjc/documents/standards/r3-reports/r3_23_antimicrobial_stewardship_amb_6_14_19_final2.pdf. Accessed 14 June 2021.
14. Stenehjem E, Hyun DY, Septimus E, et al. Antibiotic stewardship in small hospitals: barriers and potential solutions. Clin Infect Dis. 2017;65:691–6.
15. Davidson LE, Goodson K, Bear S, et al. Implementation and success of a multicenter antimicrobial stewardship network. Abstract 1677. ID Week 2016; 28 Oct 2016; New Orleans, LA.
16. Williamson JE, Logan AY, Boger MS, et al. If symptoms aren't described, antibiotics aren't prescribed: implementation of a multifaceted toolkit targeting overtreatment of asymptomatic bacteriuria across a large health-system. Open Forum Infect Dis. 2018;5:S101.
17. Isip J, Medaris LA, McCurdy L, et al. Successful use of telemedicine vs. on-site infectious diseases consultation after implementation of a system-wide antimicrobial stewardship-led staphylococcus aureus bacteremia care bundle. Open Forum Infect Dis. 2018;5:S534.
18. Roshdy D, McCarter M, Meredith J, et al. Implementation of a comprehensive intervention focused on hospitalized patients with HIV by an existing antimicrobial stewardship program: successes and lessons learned. Ther Adv Infectious Dis. 2021;8:1–8.
19. Roshdy D, Isip J, Hammer K, et al. Impact of a novel pharmacy service on antimicrobial utilization and medication-related errors in immunocompromised patient populations. Poster presented at annual atrium health quality and patient experience sharing day; August 2019, Charlotte, NC.
20. Meredith J, Roshdy D, Jaffa RK, et al. Bridging the divide between antimicrobial stewardship and surgical services: successful use of handshake stewardship with hepato-pancreato-biliary surgical services in adult patients at a large academic medical center. Open Forum Infect Dis. 2020;7(S1):S103.
21. Schmidt ML, Spencer MD, Davidson LE. Patient, provider and practice characteristics associated with inappropriate antimicrobial prescribing in ambulatory practices. Infect Control Hosp Epidemiol. 2018;39:307–15.
22. Gentry E, Sweeney C, Pion S, et al. Expanding antimicrobial stewardship into the community. ID Week 2018, Abstract 1839. Open Forum Infect Dis. 2018;5:S525.
23. Gentry E, Davis S, Collins C, et al. Development of an innovative antibiotic prescribing dashboard to enhance antimicrobial stewardship in the ambulatory care setting. ID Week 2018, Abstract 1840. Open Forum Inf Dis. 2018;5:S525.
24. Ascension – Sites of Care. https://ascension.org/Our-Work/Healthcare/Sites-of-Care. Accessed 4 June 2021.
25. Daragjati F, Guharoy R, Groves C, et al. What antimicrobial stewardship interventions are we focusing on? The experience of a large healthcare system. Open Forum Infect Dis. 2016;3:229.
26. Guharoy R, Daragjati F, Groves C, et al. Impact of a system-wide adoption of CDC core elements on antimicrobial use and Clostridium difficile in a large health system. Open Forum Infect Dis. 2017;4:S482–3.

27. Guharoy R, Fakih M, Seggerman J, et al. Universal implementation of antimicrobial stewardship programs leads to reduction in antibiotic use and cost savings: pilot results from a large United States health system. Open Forum Infect Dis. 2016;3:252.
28. Guharoy R, Fakih M, Seggerman J, et al. Impact of a dashboard to improve antimicrobial stewardship in the largest non-for-profit healthcare system in the United States. Open Forum Inf Dis. 2015;2:1411.
29. Guharoy R, Fakih M, Mends D, et al. Does one size matter? Systemwide variation of antimicrobial stewardship at Ascension Health. Abstract 1021. ID Week 2013; 4 Oct 2013, San Francisco, CA.
30. Guharoy R. Clinical decision support and antimicrobial stewardship: impact on the largest not-for-profit health system in the United States. Abstract 194260. 2016 ASHP Midyear Meeting; 5 Dec 2016, Las Vegas, NV.
31. https://intermountainhealthcare.org/about/who-we-are/fast-facts/ Accessed 5 June 2021.
32. Stenehjem E, Hersh AL, Buckel WR, et al. Impact of implementing antibiotic stewardship programs in 15 small hospitals: a cluster-randomized intervention. Clin Infect Dis. 2018;67:525–32.
33. Steneheim E, Wallin A, Fleming-Dutra K, et al. Antibiotic prescribing variability in a large urgent care network: a new target for outpatient stewardship. Clin Infect Dis. 2020;70(8):1781–7.
34. Brunisholz K, Stenehjem E, Hersh A, et al. Antibiotic prescribing in urgent care: implementing evidence-based medicine in a rapidly emerging health care delivery setting. Qual Manag Health Care. 2020;29(1):46–7.
35. Kaiser Permanente – Our Story. https://about.kaiserpermanente.org/who-we-are/fast-facts. Accessed 4 June 2021.
36. Permanente medicine: the importance of physician leadership in an integrated care system. https://permanente.org/wp-content/uploads/2017/11/Fact-Sheet_A-model-for-health-care_hyperlinks.pdf. Accessed 4 May 2019.
37. Sparks A, Fuller S, Brown D, et al. Antibiotics for upper respiratory infections. Less is more: choosing wisely at KPWA. 2017 National Quality Conference; 6 June 2017.
38. Yu K, Morcos M, Nomura J, et al. Evaluation of dedicated infectious diseases pharmacists on antimicrobial stewardship teams. Am J Health Syst Pharm. 2014;71:1019–28.
39. Sharp AL, Hu YR, Shen E, e t al. Improving antibiotic stewardship: a stepped-wedge cluster randomized trial. Am J Managed Care. 2017;23:e360–5.
40. Veterans Health Administration. https://www.va.gov/health/aboutvha.asp. Accessed 4 June 2021.
41. Kelly AA, Jones MM, Echevarria KL, et al. A report of the veterans health administration national antimicrobial stewardship initiative. Infect Control Hosp Epidemiol. 2017;38:513–20.
42. VHA Directive 1031: Antimicrobial Stewardship Programs. http://www.va.gov/vhapublications/ViewPublication.asp?pub_ID=2964. 2014. Accessed 26 April 2019.
43. Evans ME, Kralovic SM, Simbartl LA, et al. Effect of a clostridium difficile infection prevention initiative in veterans affairs acute-care facilities. Infect Control Hosp Epidemiol. 2016;37:720–2.
44. Stevenson LD, Banks RE, Strycek KC, et al. A pilot study using telehealth to implement antimicrobial stewardship at two rural veterans affairs medical centers. Infect Control Hosp Epidemiol. 2018;39:1163–9.
45. Jump RL, Olds DM, Seifi N, et al. Effective antimicrobial stewardship in a long-term care facility through an infectious disease consultation service: keeping a LID on antibiotic use. Infect Control Hosp Epidemiol. 2015;33:1185–92.
46. Graber CJ, Jones MM, Chou AF, et al. Association of inpatient antimicrobial utilization measures with antimicrobial stewardship activities and facility characteristics of veterans affairs medical centers. J Hosp Med. 2017;12:301–9.
47. Michigan Antibiotic Resistance Reduction Coalition. http://mi-marr.org/about-marr.php. Accessed 4 June 2021.
48. Mack MR, Rohde JM, Jacobsen D, et al. Engaging hospitalists in antimicrobial stewardship: lessons from a multihospital collaborative. J Hosp Med. 2016;11:576–80.

49. Vaughn VM, Ghandi T, Conlon A, et al. The association of antibiotic stewardship with fluoroquinolone prescribing in Michigan hospitals: a multi-hospital cohort study. Clin Infect Dis. 2019;69(8):1269–77.
50. Collins CD, Miller DE, Kenney RM, et al. The state of antimicrobial stewardship in Michigan: results of a statewide survey on antimicrobial stewardship efforts in acute care hospitals. Hosp Pharm. 2015;50:180–4.
51. Vizient. https://www.vizientinc.com/what-we-do. Accessed 6 June 2021.
52. Vizient. https://www.vizientinc.com/members. Accessed 6 June 2021.
53. Barnard S, Pakyz A, Kuper K, et al. How ready are hospitals to meet the Centers for Medicare and Medicaid antibiotic stewardship requirements? Abstract 968. ID Week 2016; 28 Oct 2016; New Orleans, LA.

Index

S. Doron, M. Campion (eds.), *Antimicrobial Stewardship in Non-Traditional Settings*, https://doi.org/10.1007/978-3-031-21710-4